Cambridge Elements ≡

Elements in Decision Theory and Philosophy
edited by
Martin Peterson
Texas A&M University

BEYOND UNCERTAINTY

Reasoning with Unknown Possibilities

Katie Steele
Australian National University

H. Orri Stefánsson
Stockholm University

CAMBRIDGE
UNIVERSITY PRESS

University Printing House, Cambridge CB2 8BS, United Kingdom

One Liberty Plaza, 20th Floor, New York, NY 10006, USA

477 Williamstown Road, Port Melbourne, VIC 3207, Australia

314–321, 3rd Floor, Plot 3, Splendor Forum, Jasola District Centre,
New Delhi – 110025, India

103 Penang Road, #05–06/07, Visioncrest Commercial, Singapore 238467

Cambridge University Press is part of the University of Cambridge.

It furthers the University's mission by disseminating knowledge in the pursuit of
education, learning, and research at the highest international levels of excellence.

www.cambridge.org
Information on this title: www.cambridge.org/9781108713511
DOI: 10.1017/9781108582230

First published 2021

A catalogue record for this publication is available from the British Library.

ISBN 978-1-108-71351-1 Paperback
ISSN 2517-4827 (online)
ISSN 2517-4819 (print)

Beyond Uncertainty

Reasoning with Unknown Possibilities

Elements in Decision Theory and Philosophy

DOI: 10.1017/9781108582230
First published online: August 2021

Katie Steele
Australian National University

H. Orri Stefánsson
Stockholm University
Author for correspondence: H. Orri Stefásson,
orri.stefansson@philosophy.su.se

Abstract: The main aim of this Element is to introduce the topic of limited awareness, and changes in awareness, to those interested in the philosophy of decision-making and uncertain reasoning. While it has long been of interest to economists and computer scientists, this topic has only recently been subject to philosophical investigation. At first sight limited awareness seems to evade any systematic treatment: it is beyond the uncertainty that can be managed. After all, an agent has no control over what contingencies he or she is and is not aware of at a given time, and any awareness growth takes him or her by surprise. On the other hand, agents apparently learn to identify the situations in which they are more and less likely to experience limited awareness and subsequent awareness growth. How can these two sides be reconciled? That is the puzzle we confront in this Element.

Keywords: uncertainty, decision-making, unawareness, Bayesianism, rationality

ISBNs: 9781108713511 (PB), 9781108582230 (OC)
ISSNs: 2517-4827 (online), 2517-4819 (print)

Contents

Preface

When we reason about what to do we try to include everything that we think might affect the outcome of our decision. When choosing between different career options, for instance, we may take into account things such as the earnings potential, work hours, prestige and social benefit of the career options. But often we fail to include something in our deliberation, even though it could affect the outcome of our decision in an important way. It is this limited appreciation of the full scope of relevant possibilities – dubbed *limited awareness* – that is the topic of this Element. Navigating limited awareness is a pervasive aspect of our reasoning, and yet it has hitherto been relatively little studied.

The most dramatic cases of limited awareness are when we simply lack the conceptual resources to entertain the possibilities in question. For instance, there may be features of our solar system – not captured by our best scientific theories – that bear on whether or not to pursue a career as an astronaut that even the most educated person could not entertain. To give a historical example, when early industrialists reasoned about their actions, they failed to account for the possibility of a 'greenhouse effect' at the global level; this was well beyond their scientific comprehension of the world at the time. (In this historical case, the awareness gap was eventually filled, but there may be some awareness gaps that humanity will never fill.)

Other cases of limited awareness are more mundane. Sometimes we fail to account for relevant possibilities in our decision-making due to a momentary lack of perspective – a failure to consider some otherwise familiar contingency that bears on the decision at hand. For instance, we might reason impeccably about career choices on the basis of a select set of career characteristics but overlook other characteristics, say concerning health, that for some reason happen to be inaccessible to us at the time.

Given that it covers both of the above types of cases, the category of 'limited awareness', as used in this Element, is broad. That is, it covers both dramatic and more mundane sorts of conceptual inaccessibility. This may not be the only way to conceive of an agent's awareness and the limitations thereof; but it is one that suits the context of decision-making, which is the focus of this Element. (Arguably, any study of an agent's belief state and associated awareness must make reference to some function or role that the beliefs play, whether in decision-making or otherwise.)

To be more precise, according to our use of 'awareness', an agent counts as being aware of a possibility in a decision situation just in case he or she could in that situation, without either further evidence gathering or reflection, factor the possibility into the decision. The reasons why an agent cannot factor a possibility into a decision in a particular situation can therefore be anything from limited conceptual resources to mere absent-mindedness (as one might call it). Indeed, the line between these, as far as decision-making is concerned, is far from sharp.

The main aim of this Element is to introduce the topic of limited awareness, and changes in awareness, to those interested in the philosophy of decision-making and uncertain reasoning. While it has long been of interest to economists and computer scientists, this topic has only recently been subject to philosophical investigation. At first sight limited awareness seems to evade any systematic treatment: it is *beyond the uncertainty* that can be managed. After all, an agent has no control over what contingencies she is and is not aware of at a given time, and any awareness growth takes her by surprise, at least in the sense that she can never predict *what* she might become aware of. On the other hand, agents apparently learn to identify the situations in which they are more and less likely to experience limited awareness and subsequent awareness growth. In other words, agents can predict *that* they will become more aware. How can these two sides be reconciled? That is the puzzle we confront in this Element.

We propose a way of conceiving limited awareness that does justice to its elusive character. While we build on earlier work of others, our analysis departs from this previous work in various ways. We accept that awareness growth can have radical and unpredictable effects on an agent's beliefs. But we argue that this does not preclude *anticipating* awareness growth. Moreover, we argue that, unlike the effects of experiencing 'unexpected' awareness growth, the effects of anticipating awareness growth both are quite predictable and can be captured without too radical a departure from the standard (Bayesian) model of rational preference and belief.

1 Introduction

1.1 Roadmaps to the Unknown

This Element is about our plight as reasoning agents in the world – that is, our plight as agents who seek to understand the world and how we can change it to best align with our ends. This requires some ingenuity because our perspective on the world is inherently limited. Think of it this way: our experience is confined to a more or less tiny patch of the world's history, so we can be certain of relatively little. The best we can do is try to account for all the contingencies – that is, all the ways the world *might* be – in at least as much detail as is relevant for our purposes. In this way we can build ourselves a roadmap, so to speak, for navigating the unknown.

Consider, for instance, the reasoning of a single-minded conservationist who cares only about eradicating weeds and pests. At a particular juncture, our conservationist deems that she has a limited set of options: she can release a moth that will hopefully eat the non-native cactus plant known as 'prickly pear', or she can continue with the status quo, whereby all resources are devoted to manually uprooting the pear. Our conservationist judges that which of the two options will best realise her ends depends on whether or not the world is such that the moth will eat (and kill) the prickly pear if released, and this she is unsure about. That is, these are the two possible states of the world that the conservationist deems relevant to her decision.

The conservationist's problem is summarised in Table 1, where the columns represent the *states* of the world and the rows represent the available *options*; the interior cells of the table depict the *outcomes* in each state of choosing each option. So, depending on how the world turns out, or on the true state of the world, the options yield different outcomes. In this case, there are trade-offs between the options across the states: releasing the moth is better if the first state is true while the status quo is better if the second state is true.

The prickly pear decision is a highly stylised one, but it exemplifies the general predicament we reasoning agents face, day in and day out. We are condemned to live as gamblers. By our own lights, our choices are nearly always risky ventures – we are not assured that the world will turn out one way or another, and thus whether our ends will be served more or less well by any given choice of option. Not only are we limited by our practical circumstances – the options we have to change the world – but we are limited also by our epistemic circumstances – the ability we have to discern what is true of the world and thus which of our options serve us best and even what our options *are*.

To say that we live as gamblers in fact understates the precariousness of our position in the world as reasoners. For one thing, the gambles we face in the

Table 1 A simple roadmap for navigating the unknown, a.k.a. a decision model

	moth eats pear if released	**moth fails to eat pear if released**
Release moth	pear eradicated	pear thriving, wasted resources
Status quo	pear thriving	pear thriving

pursuit of our ends are not like games of roulette or dice for which the probability of the outcomes is typically thought to be objective and easy to calculate. (We'll briefly return to this point a little later; it has been much studied elsewhere.) The other thing – the main topic of this Element – is that we typically do not have a good grasp of what *are* all the possible outcomes or contingencies that are relevant to the decision at hand. That is, it is not just that we confront the world not knowing which of the possible states of the world is actual: we do not even know what are the pertinent possibilities to begin with. Throughout the Element we refer to this latter predicament as *limited awareness*. Another form of limited awareness that will occasionally come up in this Element concerns the options available to an agent. In addition to not knowing which amongst the options she considers will best serve her ends, an agent may often not even know what *are* the options available to her.

The twenty-first-century reader may indeed have been struck by the limited awareness of our conservationist introduced in Table 1. This stylised example is in fact inspired by a historical episode in environmental management in Australia. (The prickly pear episode in Australia in the 1920s had a happy ending, as the moth that was released did in fact eradicate the highly invasive cactus. But there were other prominent cases of introduced biological pest controls in Australia that did not end so well.) With the benefit of hindsight, we can see that the conservationist failed to appreciate the complexity of decisions to introduce a biological pest control; she failed to consider other contingencies that were relevant to her decision, such as that the introduced moth might itself become a pest, eating native plant species instead of the target cactus plant. In addition, modern conservationists may see that our simple-minded conservationist failed to consider other viable options – say, a targeted chemical pest control or an alternative biological pest control – which might have served her ends at a lower risk to the ecosystem.

Let us give another stylised historical example that also highlights – and perhaps even more clearly – limited awareness due to the body of scientific knowledge available to the relevant decision-makers at the time. Although the possibility that human activity could change the climate through the 'greenhouse effect' had been discovered during the nineteenth century, it was only in the 1970s that it became relatively widely known that greenhouse gas emissions were wreaking havoc on our planet. Thus, when the first hydropower plant was built in Iceland in 1904, the country's contribution to climate change did not figure in the reasoning of the country's decision-makers. At the time, coal was the most common energy source in Europe, and importing coal instead seemed to some to be a viable alternative to building hydropower plants. Today, about 55 per cent of Iceland's energy consumption comes from hydropower and only about 2 per cent comes from coal, compared to a global average of 6 per cent from hydropower and 25 per cent from coal. So, given the pressing need to tackle the climate crisis – and since the climate impact of hydropower is generally much lower than that of burning coal – the decision to invest in hydropower was arguably right. Nevertheless, today we see that the decision was not based on all the best reasons; after all, the decision-makers in question were unaware of one of the best reasons for choosing hydropower over coal. In addition, these decision-makers were, we can safely assume, unaware of some of the options for generating significant amounts of energy that we are aware of today, such as those harnessing wind and solar power.

Examples from history allow us to witness limited awareness and subsequent growth in awareness. But fast forward now to the present. A little reflection suggests that limited awareness is not something that we reasoners have overcome. We continue to face novel scenarios and have our own forms of limited awareness. An example we focus on (in Section 6) is solar radiation management. This is a technique that could reduce (and perhaps even revert) climate change. But even its proponents admit that predicting the consequences of adopting solar radiation management on a global scale goes beyond today's scientific knowledge. Or consider carbon capture and storage. This is a relatively new technology, which consists in capturing carbon dioxide in the atmosphere and typically storing it underground, and which could be critical in combating the climate crisis. However, important uncertainties remain – for instance, about the impact of long-term underground storage of the captured carbon. Again, since this is a rather new and radical technology, the history of technological innovation would seem to suggest that it could result in consequences of which we are currently unaware.

It is these types of trying epistemic circumstances – the unavoidably parochial view of the world held by agents ranging from private citizens to individual public servants to the global community – that is the topic of this Element. The examples we will appeal to include dramatic cases of limited awareness and subsequent growth that involve novel combinations of concepts (e.g., a 'greenhouse effect' at the global level prior to the twentieth century) or even novel concepts simpliciter (e.g., an 'electron' prior to the late nineteenth century). But we will also appeal to more mundane cases of limited awareness and subsequent growth due to temporary shifts in attention or imaginative ability. The plight of the conservationist, an example we will pursue in this introductory section, arguably lies somewhere in the middle of the spectrum.

There are two things to say about our liberal stance on what counts as limited awareness that may help orient the reader from the outset as to the target of our inquiry and our approach. The first is that, as the reader may well discern, we examine limited awareness in a practical decision-making context (and as such we will be appealing to and extending the tools of 'Bayesian decision theory'). While we focus on decision-making, we have little to say about an agent's basic values or ends; we simply take them as given. (That is, we leave the analysis of values or ends for others to address.) Nonetheless these ends have an important bearing on our epistemological project. It is not just that we are ultimately interested in how an agent reasons about what to do to further her ends. Our very understanding of her epistemic state and associated (limited) awareness is intimately tied to the pursuit of her ends.

Moreover, we doubt whether an agent's epistemic state and its limitations can even be well understood in the absence of *some* functional role that the epistemic state plays. The functional role that we are interested in is decision-making, but we allow that others may have different projects in mind and may thus understand an agent's epistemic state and her limited awareness in different ways that are moreover less liberal about what is genuine limited awareness and what is a mere mistake. That is, although we think that limited awareness is an important phenomenon that may be explored in a range of guises, by appeal to a variety of models, our particular interest is limited awareness in the context of decision-making. As such, we deem limited awareness to be concerned with whatever is the agent's decision frame at the time. In particular, we take an agent to be aware of a possibility, in a given decision situation, just in case she could in that situation – without either further evidence gathering or, say, overcoming any defects in imagination – factor the possibility into her decision. Her awareness may in this sense be limited and subject to growth.

Our conservationist, for instance, may well have many (at least implicit) ideas about the way the world is, including the weather, her family and friends, and so on. But these ideas are in a sense idle, at least in the context of her current options and ends, which concern the eradication of pests. Generally speaking, there may or may not be a richer story to tell about an agent's epistemic life. This Element, however, aims only to capture a part of this story. When we talk of an agent's epistemic perspective, we mean her current views about the possible contingencies, or ways the world might be, in so far as those contingencies play a role in her reasoning about what to do *now* to further her ends. In other words, an agent's epistemic perspective is relative to a decision problem, as we model it.

The second thing to note is that, while we thus seem to engage with rather hapless agents – including agents that may seem quite far from any ideal state of awareness, even accounting for the limitations of the best science of the day – our project remains normative. We will examine how an agent *should*, rationally speaking, navigate her limited awareness and awareness growth. In particular, we take as our starting point agents' differing degrees of awareness – and, as we shall later see, the different extent to which agents are aware of their unawareness – and we ask what principles of rationality such agents should satisfy, for instance, when their awareness grows or when they simply predict awareness growth. So, the Element seeks to answer normative questions about agents who are less than epistemically ideal, in that they lack full awareness.

Now, any normative project of this kind will inevitably to some extent be prescriptive. That is, the principles of rationality we discuss are useful not just for *assessing* the rationality of agents but also for *guiding their deliberations*. But our primary aim is the former rather than the latter; we do not set out to offer principles that it would necessarily be wise (or even possible) to *apply* whenever one finds oneself in, say, a state of growing awareness or anticipated awareness growth. Still, we hope that by, for instance, revealing what principles one should ideally satisfy in situations of limited unawareness – and, moreover, illuminating the nature of (un)awareness and the situations where people have previously been demonstrably unaware – this Element can help readers make better decisions and reach more justified conclusions when they find themselves in such situations.

1.2 Internal Consistency and Its Limits

Since we will investigate limited awareness in the context of an agent's decision-making, we will appeal to standard (Bayesian) decision theory as our starting point. In the remainder of this introductory section we will explain how we will build upon standard decision theory – why it does not accommodate limited awareness and what we will seek to fill in, in the Element.

The standard decision-theoretic account of our reasoning goes beyond simple roadmaps such as those we have described. The roadmaps that the theory offers account not only for the supposed possible contingencies or ways the world might be but also, typically, their relative plausibility. To be *rational*, i.e., to reason well, one's judgements of relative plausibility must be *internally consistent*. Another requirement of rationality is that one's judgements of relative desirability be internally consistent. Indeed, decision theory can be understood as a theory of internal consistency. It tells how our epistemic and evaluative judgements or attitudes must 'hang together' so as to yield clear choices of action that are not self-defeating with respect to our ends.

This Element is about the limits of internal consistency, in particular due to an agent's *(limited) awareness*, or what she perceives to be the possible contingencies or ways the world might be. But we need an understanding of the guidance that internal consistency can provide in order to see what are the shortcomings of this guidance. In what follows, we start by artic- ulating the guidance (1.2.1) before looking more closely at how arguments from internal consistency work (1.2.2) and what are their inherent limitations (1.2.3). One way to understand the point of this Element is to consider that we want to go beyond the uncertainty that decision theory typically deals with – that is, beyond the type of uncertainty that can be treated normatively in terms of internal consistency. However, we acknowledge that we are only taking one step beyond this uncertainty and that further steps may have to be taken.

1.2.1 Introducing Probabilities

We said that agents consider the relative plausibility of the possible ways the world might be. Put differently, agents have varying *degrees of confidence* – also known as *degrees of belief*, or as *credences*, which is the term we shall mostly use – in ways the world might be. It is as if they weigh the competing possibilities on a scale with multiple arms. The common wisdom is that, as an arm gets more weight, the others should collectively get less weight. To be more precise: credences are rational only if they can be represented as probabilities. This norm is often referred to as *probabilism*. For instance, if our conserva- tionist assigns much weight, or has relatively high credence, say, of 0.9, in the moth eating the pear if released, then on pain of inconsistency she must assign little weight, or have relatively low credence, here 0.1, in the moth *not* eating the pear if released.

Let us more thoroughly describe our conservationist's credences, as perti- nent to the choice problem depicted in Table 1. As noted, what matters for determining how well her options realise her ends is whether the released moth will eat the prickly pear or not – that is, which of these states of the world is actual – which we can denote M and $\neg M$ respectively. Strictly speaking, our conservationist is also unsure about what she will do, whether she will release the moth or not, denoted R and $\neg R$ respectively. This yields four rel- evant possibilities for how the world might be: $R\&M$, $R\&\neg M$, $\neg R\&M$ and $\neg R\&\neg M$. We assume that our conservationist has credences in each of these fine-grained possibilities or *outcomes* that are each non-negative and together sum to one. Her credences in all other propositions involving M and R can be derived in conformity with the probability calculus. Moreover, presumably our conservationist's credences in *M* versus ¬*M* do not depend on her credences in *R* versus ¬*R*. That is, $P(M|R) = P(M|\neg R)$, where *P* represents the agent's credences and $P(M|R)$ denotes her *conditional* credence in *M* given *R*. That is, in this case we have *act-state probabilistic independence*, but this need not always be so.[1]

Table 1 is the most economical depiction of our conservationist's choice problem. But note that the view of the world she brings to bear on this choice problem, and her associated credences, may be somewhat more complicated. Perhaps she entertains other potential properties of the world in an effort to form judgements about the relevant states of the world. For instance, perhaps our conservationist recognises that there may or may not be a drought during the year following the potential release of the moth, denoted *D* and ¬*D* respectively. She does not care about droughts. Our assumption is that she cares only about the eradication of weeds and pests. So in a sense whether or not there is a drought does not matter to her. Nonetheless, the consideration of whether there will be a drought may assist our conservationist in forming her credences in *M* and ¬*M*. After all, by the law of total probability, $P(M) = P(M\&D) + P(M\&\neg D)$. Plausibly, our conservationist arrives at a settled credence in *M* by considering her 'component' credences in $P(M\&D)$ and $P(M\&\neg D)$. This is to say that our conservationist's roadmap may look more like Table 2.

In general, there is a *space of propositions describing ways the world could be* about which the agent has an opinion that bears on her practical reasoning at some given time. This space of propositions about which she has an opinion is assumed to have a certain completeness in structure. In technical terms, it forms an algebra $\mathcal{F}$ with the following characteristics (which means that it is what is called a *Boolean* algebra):

- $\mathcal{F}$ contains a contradictory proposition ($\bot$).
- $\mathcal{F}$ contains a tautologous proposition ($\top$).
- $\mathcal{F}$ is closed under disjunction, conjunction and negation. That is, if *A* and *B* are in $\mathcal{F}$, then $A \lor B$, $A\&B$ and ¬*A* and ¬*B* are also in $\mathcal{F}$.

---

[1] The knowledgeable reader may discern that our presentation of the agent's decision model follows that of Jeffrey (1965), as opposed to Savage (1954).

Table 2 A more detailed roadmap for navigating the unknown

	moth eats pear; drought	moth eats pear; no drought	moth fails to eat pear; drought	moth fails to eat pear; no drought
Release moth	pear eradicated	pear eradicated	pear thriving, wasted resources	pear thriving, wasted resources
Status quo	pear thriving	pear thriving	pear thriving	pear thriving

Table 3 A yet more detailed roadmap for navigating the unknown

	moth eats pear; drought; pest	moth eats pear; drought; no pest	moth eats pear; no drought; pest	moth eats pear; no drought; no pest	
Release moth	pear eradicated, pest	pear eradicated, no pest	pear eradicated, pest	pear eradicated, no pest	...
Status quo	pear thriving	pear thriving	pear thriving	pear thriving	...

The rational agent has credences in the propositions in $\mathcal{F}$ that can be represented by a probability function P. That is, $P(A) \in [0, 1]$ for all A in $\mathcal{F}$; $P(\bot) = 0$; $P(\top) = 1$; $P(A \vee B) = P(A) + P(B)$ for all mutually exclusive A and B in $\mathcal{F}$.

1.2.2 Rationality as Internal Consistency

Why think that rational credences are probabilities? There are various arguments for this position. A relatively straightforward one is known as the 'Dutch book argument'.[2] It turns on the claim that an agent's credences are effectively her 'betting odds' or the proportion of the stakes she'd be willing to pay for a bet that yields the stakes if the proposition in question turns out true but yields nothing otherwise. It is shown that if and only if her betting odds over the space of propositions conform to the probability calculus, the agent is *not* vulnerable to accepting a set of bets that would guarantee her a sure loss

---

[2] An argument like this was first suggested by Ramsey (1926). For a recent overview of different Dutch book arguments, see Richard Pettigrew's Element in this series (Pettigrew 2020).

(measured in monetary terms). Positioning oneself for a sure loss is considered a marker of inconsistency, albeit of a *pragmatic* kind. So one's credences had better be probabilities. Note that other arguments for credences being probabilities turn on inconsistencies of a *non-pragmatic* kind. For instance, credences that do not conform to the probability calculus are shown to be *accuracy dominated* in the sense that some alternative probabilistic credence function would be more accurate (roughly, closer to the truth) no matter how the world turns out.[3]

We have been emphasising the role of rational credences in deliberating about what to do. Indeed, the standard wisdom is that probabilistic credences, together with a cardinal value or *utility* function over outcomes (in our example: the cell entries in Table 1) – which represents how much the agent in question values the outcomes – determine the *expected utility* of risky options, which is the basis for their relative desirability and thus choice-worthiness. The expected utility of an option (or indeed any prospect or claim about the world represented by a proposition) is the sum of the probability of each possible way in which the option or prospect may be true multiplied by the utility of that way it may be true. The higher the expected utility, the better, according to expected utility theory.

To make the above more concrete, recall the choice problem of our conservationist, as represented by Table 1. The conservationist is considering two options: release the moth (R), and don't release the moth ($\neg R$). There are only two states of the world that she considers relevant to the outcome of her options: the moth eats the pear if released (M), or it does not ($\neg M$). Now let U be the conservationist's utility function over outcomes. Then the expected utility (EU) of the conservationist's options, according to her, are given by:

$$EU(R) = U(R\&M)P(M \mid R) + U(R\&\neg M)P(\neg M \mid R)$$
$$EU(\neg R) = U(\neg R\&M)P(M \mid \neg R) + U(\neg R\&\neg M)P(\neg M \mid \neg R).$$

By the aforementioned assumption that act-state probabilistic independence holds in this case, the above equations reduce to:

$$EU(R) = U(R\&M)P(M) + U(R\&\neg M)P(\neg M)$$
$$EU(\neg R) = U(\neg R\&M)P(M) + U(\neg R\&\neg M)P(\neg M).$$

There are again various arguments for why one ought to evaluate and rank risky options according to their expected utility.[4] (Call this the

---

[3] See, e.g., Joyce (1998).
[4] For overviews of these arguments, see Briggs (2017) and Steele and Stefánsson (2015).

expected utility principle.) One kind of argument appeals to the infinite long run: expected utility matches what one would be more or less sure to gain were the choice repeated over and over (provided certain independence conditions hold between the individual decisions). The more prominent kind of argument is known as the *expected utility representation theorem*, which is all-encompassing in that it supposedly justifies rational credences being probabilities together with the expected utility principle, all in one hit. There are several different versions of this theorem, but they have a similar form. They appeal to consistency in the ranking of options comprising a specially engineered rich set of options. (Note that the 'ranking' of options here means how they are ordered in terms of the agent's judgement of their relative desirability or 'choice-worthiness'.) One consistency requirement, for instance, is *transitivity*, which requires that if an agent ranks option A over B and B over C, then she ranks A over C. In short, the expected utility theorem is the result that if and only if an agent's ranking of the relevant options satisfies a number of consistency constraints, including transitivity, she can be represented as having credences measured by a probability function and judgements of relative desirability measured by a cardinal function that conforms with the expected-utility principle.[5]

What we have just presented is the orthodox position in decision theory. Indeed, the arguments for rational credences being probabilities and for rational evaluations of options satisfying the expected utility principle are core results in decision theory. This is not to say, however, that these arguments have not been challenged. Different (generally weaker) constraints on what counts as rational credence have been fruitfully explored, where 'rational' is still understood in terms of internal consistency.[6] And different (again, generally weaker) constraints on rational evaluations of options have also been fruitfully explored, where, again, 'rational' is understood as internal consistency.[7] We do not pursue these debates in this Element, however. Rather, we stick with the orthodoxy, at least to the extent that it is applicable. But this should not be interpreted as strong endorsement of expected utility theory. We build on the orthodox theory for reasons of simplicity. The issues raised in this Element concern

---

[5] For some classical representation theorems, see Ramsey (1926), Savage (1954) and Bolker (1967).

[6] In particular, that rational credences need not be *precise* is a popular view. For an overview, see Bradley (2019).

[7] Lara Buchak (2013) has developed an influential theory with weaker constraints on the evaluation of options than those in orthodox decision theory. Stefánsson and Bradley (2019) criticise Buchak's theory and defend an alternative view.

the limits of internal consistency and, as such, are orthogonal to the debate between expected utility theory and alternative formalisations of internal consistency.

1.2.3 Beyond Internal Consistency

So internal consistency provides standards for our roadmaps. But one might have further concerns about whether an exemplary roadmap, in this respect, will really be a good guide to the world. What if the roadmap is not very faithful to the world, despite being internally consistent? There are two issues that might come to mind. The first is that the agent's credences may not be very sensible. After all, we would not take very seriously an agent who is extremely confident that the moon is made of green cheese, even if the agent's credence function overall conformed to the probability calculus.

The other issue is that the agent may have *limited awareness* of the various contingencies, or possible ways the world might be, such that her perception of her own decision problem, and what matters, is rather narrow. For instance, perhaps our conservationist simply does not take account of whether or not there is a drought, and so her credences lack the sophistication that would come from accounting for this contingency (as per Table 2 rather than Table 1). Worse still perhaps, she might not recognise contingencies that have a more obvious bearing on her evaluation of outcomes. For instance, our conservationist might not realise that the released moth could possibly eat native plant species, thereby itself becoming a pest. Table 3 includes this contingency in addition to the drought contingency that was already introduced in Table 2.[8] (To save space, only some of the states and outcomes in Table 3 are explicitly labelled.) In addition, and as previously discussed, our conservationist might not be aware of all the options that are available to her.

This Element is concerned with the problem of limited awareness. We do not try to address the problem of nonsensical credences. For what it's worth, we do not think that the two issues are entirely unrelated, since the more possible ways in which the world might be that an agent is aware of, the more checks and balances there are on her credences. For instance, if one is to maintain high credence in the moon being made of green cheese, without there being a conflict within one's overall credal state, then one needs to also maintain some other odd credences, like that telescopes are generally misleading, at least when it comes to the composition of the moon. And so on. There may be no substantial norms governing the credences that an agent has at any particular

---

[8] Table 3 is what we later call a *refinement* of the agent's possibility space described in Table 1.

point in time, beyond the probability calculus. But we leave that as an open question.

1.3 Limited Awareness in Perspective

So our focus is on how an agent's roadmap to the unknown may fail to capture all the important contingencies or ways the world might be. Standard decision theory is silent on how to respond to this failure. The primary reason for this is that standard decision theory does not acknowledge the failure! That is, the theory, as it is standardly used, simply assumes that a roadmap, at least in outline, is given to a decision-maker and is only applicable once the decision-maker has such a roadmap. Another way to put this is that it is standardly assumed that the framing of a decision problem, including the relevant space of possibilities, is not part of the theory; the theory can only be applied given some framing. But a general account of reasoning should not, we think, overlook issues of framing. Instead, it should acknowledge that an agent's roadmap or framing of a decision problem is integral to her reasoning. And it should say something about whether a rational agent perceives gaps in her roadmap and how she fills in those gaps over time. In other words, it should say something about how to rationally respond to a lack of awareness and growing awareness.

To see that the move to modelling limited awareness is an important one, let us situate it in a series of developments towards a more thoroughly subjective decision theory. We are then in a better position to determine whether this is an important next move.

1.3.1 Towards a Truly Subjective Expected Utility Theory

Let us return to the development of the standard model. The history of standard expected utility theory can be understood in terms of social scientists' efforts to provide operational definitions of key reasoning attitudes like credence and desire in more and more realistic settings. We want to position our inclusion of growing awareness as the natural next step in the trajectory towards a fully general model of reasoning. To see roughly how this goes, we will describe an earlier, somewhat analogous move along the trajectory: the move from the von Neumann–Morgenstern (1947) model to that of, say, Savage (1954) or Jeffrey (1965).

Start with the expected utility theory of von Neumann and Morgenstern (vNM). The theory establishes the conditions under which an agent ranks risky options in accordance with their expected utility. But the model vNM introduce effectively has only one free variable or subjective dimension: the extent the agent desires or values various outcomes. Other aspects of the options are fixed since these are assumed to be an objective part of the decision problem that is not specific to the agent in question. In particular, the nature of the final outcomes is assumed to be an objective matter, and moreover transparent to all, as well as the probabilities by which the various options may yield these outcomes (and, by implication, the options themselves). The theory establishes that if and only if the agent's ranking of these options satisfies some proposed internal consistency constraints, then the agent's strength of desire can be measured by a cardinal utility function (unique up to positive linear transformation) such that she evaluates options according to their expected utility. (Recall our earlier remarks about the expected utility theorems.)

The vNM model offers significant insights about the structure of reasoning; in particular, it provides a powerful way to measure and thus conceptualise strength of desire. The problem is that the model is only applicable when its assumptions hold. And they rarely do hold (Hansson 2009). In most choice settings, the probabilities with which the available options yield the different outcomes are not objective parts of the decision problem but rather are something that the decision-maker brings to the decision problem – hence the development of expected utility theories that allow for more subjectivity in the characterisation of the options. The theories of Savage and Jeffrey, for instance, accommodate desires that are specific to the agent, as vNM do, but in addition allow for probabilistic credences that are specific to the agent (*subjective* desire *and* credence). The ingenuity of these theories is that they still allow a way of measuring and thus conceptualising these attitudes in terms of the agent's ranking of options (in so far as this ranking is internally consistent or rational).

The subjective expected utility theories of Savage and Jeffrey are more generally applicable than that of vNM. They propose a way of measuring and thus making sense of subjective desire and credence. But note that there remains an important 'objectivity' assumption: the options that are ranked are still in large part given to the decision-maker (in the sense of being an objective part of the decision problem) rather than being specific to her. In Savage's model, for instance, the options are functions from states of the world to outcomes. It is not that the probabilities for the states of the world are objectively given (as in the vNM model). That is where the agent's own credences come in. But the states of the world themselves and the nature of the outcomes are assumed to be objectively given or external to the agent's reasoning.

So while Savage's and Jeffrey's theories make an advance on that of von Neumann and Morgenstern, these theories too have limited applicability. In particular, they are not applicable when the very states of the world and outcomes (or more generally, the ways the world might be) cannot be assumed to be external to the agent but are rather specific to her own perspective, in particular her specific level of awareness. That is the kind of scenario we will focus on in this Element.

1.3.2 A Limit to What Can Be Modelled?

Now one might think there is nevertheless good reason for not trying to model limited awareness. To begin with, this arguably makes for one too many free variables in our model of an agent's reasoning that cannot, even in principle, be empirically settled. We lose sight of what our concepts mean in an operational (or functional) sense, and providing this meaning was one of the great advances of decision theory. The ingenuity of Savage's expected utility theory, for example, is that it offers a way to understand credence and desire in terms of choice behaviour. But there might seem to be no conceivable choice scenario that could reveal an agent's attitude to an outcome of which she is unaware without thereby making her aware of the outcome.

Secondly, one might think it is in any case pointless to model limited awareness in a normative decision model. Lack of awareness is typically not something that an agent can correct (by herself). Compare this to, say, nontransitive preferences. An agent can check for herself whether her preferences are transitive or not; and if they are not, she can try to make them transitive. By contrast, an agent normally cannot check whether she is in fact unaware of something; and even if she suspects that she is unaware of something, she cannot simply correct for this lack of awareness by becoming more aware.

These are two important challenges for the project of modelling limited awareness. We will not, however, respond to either of them directly in a satisfying way. We rather acknowledge, here at the outset, reasonable scepticism about the project we embark upon. Our hope is that the scepticism is mitigated by our treatment of limited awareness in the remainder of the Element.

Moreover, we are not alone in the quest to understand limited awareness and how it affects an agent's reasoning. A small but strong cohort of economists, computer scientists, and philosophers have already made important progress towards this goal, and we are indebted to the foundations they have laid. The very notion of limited awareness – or *un*awareness, as they typically call it[9] – is due to early work by economists and computer scientists. An extensive review

---

[9] Since awareness is typically a matter of degree – in the sense that an agent can be more or less aware (as in, aware of more or fewer possibilities) – we find the term 'limited awareness' to be more apt, in most cases, than 'unawareness'. And indeed that is the terminology we shall typically use. However, since 'unawareness' has come to be widely used, in particular by economists, for what we think should be called 'limited' awareness, we will occasionally use the term 'unawareness' for limited awareness, for instance, when discussing the works of these economists.

of this work can be found in Schipper (2014). Philosophers too have explored the challenge that growing awareness poses for the traditional probabilist models, typically under the guise of 'the problem of new theories' (Earman 1992; the problem was originally raised by Glymour 1980 as the counterpart to 'the problem of old evidence'). Richard Bradley (2017) has recently turned philosophical attention to the general problem that (un)awareness poses for rationality; his own work draws on a series of recent papers by the economists Karni and Vierø (2013, 2015, 2017).

We will expand on these earlier contributions in relevant places throughout the Element. For now, let us simply give a taste of how the challenges raised have been at least partially met by others. For instance, on the first: Piermont (2017) illustrates how we can, by observing a person's choices between what he calls 'contingent plans' (that is, conditional options of the form: if state s obtains, then choose c), tell whether or not a person anticipates her awareness to grow. In short, the idea is that a person anticipates awareness growth just in case she is willing to take on some cost to postpone a choice between contingent plans, even when all such plans that she can conceive of are available. Moreover, Karni and Vierø (2017) show that even if a person anticipates awareness growth, as long as her preferences satisfy certain consistency constraints, then she can be represented as maximizing expected utility. The representation even allows us to infer how (un)desirable the decision-maker predicts a currently unknown outcome to be.

Turning to the second challenge: although it is true that one cannot determine whether or not one is unaware of something, nor immediately become aware of that something in case one is unaware, one can take steps to increase one's level of awareness if one suspects that there is something one is unaware of. That is, one can conduct (formal or informal) experiments that can be expected to reveal contingencies that one is unaware of, if there are such contingencies. Moreover, one can make plans for how one will adjust one's attitudes, and what choices one will make, if one does become more aware. In fact, it would seem that one *should* in certain choice situations suspect that there is something that one is unaware of; hence, one arguably should, in some situations, make plans for how to respond to growing awareness.

The issues just raised – concerning how one should respond to awareness growth and the extent to which one can and should plan for such growth – will be explored in a systematic fashion throughout the Element (more on how we will proceed shortly). For the moment, the idea is just to get a feel for how (un)awareness, as amorphous as it may sound, nonetheless admits of structured treatment according to usual decision theoretic principles. That said, we are sympathetic to the general worry that there is only so much that can be said or done about limited awareness. It is just that, notwithstanding the significant and pioneering contributions to date, not enough has yet been said. Limited awareness and its perils is an interesting aspect of our epistemic predicament, and it has not yet been addressed in a fully general and comprehensive way. This Element is intended as a step towards remedying that situation.

1.4 How We Will Proceed

As noted, we will approach the problem of limited awareness in an incremental fashion. Even if it turns out that there is little to say about one's limited awareness at a time, there is plausibly a lot to say about *changes in awareness*, specifically *awareness growth*, over time. After all, our roadmaps for navigating the unknown are not finished deeds but rather works in progress. They are subject to feedback from the world itself, as events unfold and the agent comes to new realisations. That is, in any case, the starting insight for the approach taken in this Element. We will approach the issue of limited awareness – what it is and whether there is anything to be done about it – by considering first *changes* in awareness.

In fact we begin in Section 2 at an even more preliminary point. We consider, in qualitative terms, the kinds of feedback that the world may provide on our reasoning roadmaps. As said, these roadmaps are not finished deeds. Consider our conservationist. Subsequent to her deliberations, let's say she releases the moth. The story does not end there, of course. Presumably, after some time, it will be apparent to her whether or not the prickly pear has been eradicated and thus which state of the world is actual. A more thorough roadmap than those we have considered thus far would in fact seek to anticipate such learning events. That is, in addition to other sorts of properties of the world, such as whether a moth population will eradicate a prickly pear population, it may be important to also anticipate one's interactions with the world, in particular, what one will learn and when, and what options one may choose between and when. The popular format for these more thorough roadmaps is the *sequential-decision* model. Section 2 proposes a way to read such models. But no matter how thorough, a roadmap that attempts to account for all relevant future contingencies comes up against the world as time unfolds. Actual events will either be consistent with the roadmap or not.

In Section 3 we elaborate on the kind of feedback from the world that is the focus of this Element – the realisation of unfamiliar contingencies, or, in other words, the realisation of limited awareness. We articulate the different types of awareness growth that such a realisation may herald and go on to consider how such growth is best modelled. We finally reflect on how our approach to awareness growth borrows insights from the work of others across different disciplines.

Sections 4 and 5 go on to consider the impact of awareness growth on rational credences. While others who have investigated this question have a rather sanguine view of the impact of awareness growth on credences, we argue that awareness growth may have highly disruptive, far-reaching impacts, at least in some cases. We proceed to offer a characterisation of the better behaved cases: when awareness growth has a more conservative impact on one's credences.

We then turn, in Sections 6 and 7, to the question of whether there is anything to be done in advance to stave off radical changes in one's credences due to awareness growth. We argue that there is a sense in which one can and indeed *should* plan ahead for awareness growth, even if, at the end of the day, there are no assurances that awareness will not change in unforeseen ways. Encouraging decision-makers to plan for awareness growth, as well as providing them with the tools for such planning, is arguably decision theory's most important contribution to the problem of limited awareness.

Section 8 summarises the findings of the Element and suggests avenues for future research.

2 Sequential Decision Models and the Test of Time

2.1 Introduction

As we explained in the introductory section, one can view the introduction of limited awareness into decision models as the natural next stage in decision theory's historical trajectory, from the 'objective' expected utility theory of von Neumann and Morgenstern (1947) to the 'subjective' expected utility theories of Savage (1954) and Jeffrey (1965). In this section we begin the task of articulating what a more thoroughly subjective decision model that can accommodate changes in awareness looks like.

The starting point for our investigation is the experience of failure. In the course of time, it often becomes apparent that one's roadmap or decision model has failed to account for all contingencies. The recognition of failure, we suggest, is the first step to i) acknowledging limited awareness and subsequent growth in awareness and ii) reflecting on when such growth might happen again. These are the major themes to be developed later in this Element.

There are in fact several kinds of feedback that the world may provide on one's roadmap. The failure to account for all contingencies is just one kind of feedback. This section will consider the other kinds of feedback as well, by way of putting limited awareness in perspective. First, however, we need to introduce a more detailed kind of roadmap, known as a *sequential-decision*

Table 4 Conservationist's roadmap

	eats; drought	eats; no drought	no eats; drought	no eats; no drought
Release moth	eradicated	eradicated	thriving, waste	thriving, waste
Status quo	thriving	thriving	thriving	thriving

model. Such a model makes explicit an important kind of future contingency: the agent's own interactions with the world – not only when and what she will be able to *choose* but also when and what she will *learn* about the world. An important aspect of the feedback that the world provides is whether an agent learns the things about the world that she expects to learn.

This section thus proceeds, in Section 2.2, to lay out sequential-decision models. We can then identify, in Section 2.3, four major kinds of feedback the world may provide on such models. Once we have limited awareness in perspective, we can elaborate on the particular forms it may take and how it is best modelled. But that task must wait until Section 3.

2.2 Situating Oneself in the Stream of Events

Return to the plight of our single-minded conservationist. Assume, for starters, that the contingencies she is tracking (apart from her own choice of action) concern whether or not the moth eats the pear if released and whether or not there is a drought. That is, her roadmap, or personal decision model, is as per the second decision problem in the introductory section, reproduced in Table 4 in abbreviated form.

Let us now add some further detail to our conservationist's roadmap, concerning her other interactions with the world. As noted, such interactions may include i) when she receives information from the world, which we may refer to as 'learning events', and ii) when she makes a choice that impacts on the world, which we may refer to as 'choice events'.[10] Besides her choice of whether or not to release the moth, let us assume that our conservationist predicts only one relevant interaction with the world: she predicts that she will learn, prior to making her choice about whether to release the moth, whether or not there will indeed be a forthcoming drought. To keep things simple, imagine that she predicts she will somehow learn this with certainty, even though that is rather

---

[10] In game theory, the terms 'nature's moves' versus 'agents' moves' are used. This is along the lines of the distinction we are drawing here but does not quite coincide, since 'nature's moves' may or may not be learnt by the agent.

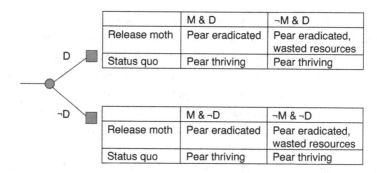

	M & D	¬M & D
Release moth	Pear eradicated	Pear eradicated, wasted resources
Status quo	Pear thriving	Pear thriving

	M & ¬D	¬M & ¬D
Release moth	Pear eradicated	Pear eradicated, wasted resources
Status quo	Pear thriving	Pear thriving

Figure 1 A roadmap as per Table 4 with predicted learning

implausible – forthcoming weather is not ordinarily the sort of thing one can learn with certainty.

Our conservationist's sequential-decision model is shown in Figure 1. As per convention, the circle nodes represent learning events, when the world presents new evidence, the possibilities for which are represented by the branches emanating from the node; the square nodes represent choice events, when the agent will have the opportunity to choose amongst options, normally also represented as branches emanating from the node. Here we abbreviate the model slightly by inserting the relevant decision table at the choice nodes. (Recall that M denotes that the moth eats the pear if released, while D denotes that there is a drought.)

As per their *static* (or *single-decision*) counterparts, sequential-decision models are *not* typically interpreted in a fully subjective way. The space of contingencies, including the agent's own interactions with the world, are taken to be given to the decision-maker – that is, they are assumed to be an objective part of the decision problem – rather than being specific to her. For instance, with respect to Figure 1, it is typically taken to be an objective part of the decision problem that the agent will learn whether or not there will be a drought (effectively assigning what she learns probability one) and will subsequently choose whether or not to release the moth. But we pursue a more thoroughly subjective interpretation of sequential-decision models. It is not merely that the agent assigns her own subjective probabilities to whether or not there will be a drought, but it is moreover her own fallible prediction that she will learn this information at a given time and then will subsequently be presented with a choice as to whether or not to release the moth. Of course, the finer details of 'subjectivising' the space of contingencies in this way will need to be ironed out. For now we ask that the reader simply go along with our reading of sequential-decision models such as that depicted in Figure 1, and our 'subjectivising' project more generally.

We make one further point before continuing: if learning and choice events are a matter of the agent's own predictions, then strictly speaking, these should feature in the space of contingencies. That is, the relevant propositions should feature in the algebra $\mathcal{F}$ over which the agent has opinions. For instance, as well as entertaining the contingencies of drought, D, and no drought, $\neg D$, Figure 1 shows that our agent also entertains learning, prior to acting, of drought, or else of no drought, which we might denote L_D and $L_{\neg D}$ respectively.

Similarly, the agent's predictions about her future credences and desires should also feature in the algebra over which she has opinions. And so too her predictions about the future choices she may encounter. But including all this in the agent's algebra would make it rather complicated. So the convention is not to include these propositions, unless for some special purpose (such as arises in Section 7). One can think of a sequential-decision model as a relatively simple way to represent all these extra predictions about how the agent will interact with the world, without expanding the agent's algebra (instead using circle and square nodes in a tree-like structure).

This abbreviated way of representing the agent's predictions about her future attitudes and interactions with the world is only accurate, however, under some important assumptions regarding how an agents sees her future self in relation to the external world. We will take these assumptions for granted throughout this Element as they are crucial for the type of methodical examination of learning and anticipated learning that we will be carrying out in this Element. Roughly speaking, it is assumed that the agent does not predict her future self to be a skittish and irrational character by her current lights. Rather, she predicts her future self to be an exemplary reasoner who moreover has credences and desires that are stable, or continuous with those she currently holds. Moreover, she predicts her future self to be properly responsive to the evidence, such that she will learn some proposition only if it is really true. This means, for instance, that our conservationist regards L_D as entailing D and $L_{\neg D}$ as entailing $\neg D$.[11] It is moreover implicit in the sequential-decision model that the agent predicts she will update her credences (and desires) upon learning in a way that accords with her rational plans.

It is generally accepted that the rational plan for updating or changing one's credences in response to learning that some proposition (in which one had positive credence) is certain, as befalls our conservationist in Figure 1, is to

---

[11]   To be clear, the assumptions concern the relationship between an agent's predictions about the external world and her predictions about her future self. They do not concern what is really true of the external world, nor do they concern what is really true about the agent's future self or selves.

conditionalise on what one has learnt. The new credence in any proposition is just the old credence, conditional on what is learnt.[12] Accordingly, our conservationist predicts that her credence in the moth eating the pear, M, in the case that she learns of a drought at the circle node, L_D (or equivalently, D), will be $P(M|L_D)$ (that is, $P(M|D)$). Likewise, she predicts that her credence in M, in the case that she learns that a drought will not occur, $L_{\neg D}$ (or equivalently, $\neg D$), will be $P(M|L_{\neg D})$ (that is, $P(M|\neg D)$). Note that for the remainder of this section we will suppress the propositions describing the agent's own learning, since we are assuming that the agent treats these propositions (for instance, L_D) as equivalent to those describing what is learnt (in this case, D).

It follows that our conservationist's current credence in M equals a weighted average of what she predicts her credence in M to be in the future, where the weighing is determined by her current credence in D vs. $\neg D$. (In other words, the agent satisfies a principle called *Reflection*, which we discuss in detail in Section 7.) To see this, note that by the law of total probability, for any M and D:

$$P(M) = P(M \mid D)P(D) + P(M \mid \neg D)P(\neg D).$$

Therefore, by our assumption that, first, our conservationist predicts that she will learn either D or $\neg D$, and, second, she predicts that her credence in M, in the case that she learns of D, will be equivalent to $P(M|D)$, and similarly for $\neg D$, it follows that her current credence in M equals what she *expects* (i.e., her weighted prediction) her credence in M to be in the future. To take a concrete numerical example, let's say that our conservationist's credences are such that $P(M|D) = 0.6$ and $P(M|\neg D) = 0.9$. Then, if we assume that $P(D) = 0.5 = P(\neg D)$, it follows from the above that $P(M) = 0.75$, and this, moreover, is her expected future credence in M.

Now it may well be that the learning of the drought is irrelevant for our conservationist's choice, because whether her credence in $P(M)$ increases from 0.75 to 0.9, or else decreases to 0.6, her preferred option is still to release the moth. (This will be so if eradicating the pear is sufficiently better than the status quo relative to how much worse the wasted resources are compared to the status quo.) Or it may be that she predicts the learning to be relevant in the sense that it will affect her choice. Either way, that is not our focus here. The point is rather to show how an agent's roadmap may be rather sophisticated in keeping track

---

[12] The rule of conditionalisation is sometimes referred to as *Bayesian learning*. Indeed, the term *Bayesianism* incorporates the norm that credences should be probabilities (probabilism) as well as the norm that credences should be revised in accordance with conditionalisation (or generalisations thereof). Hence why subjective expected utility theory, as described in Section 1, is sometimes referred to as *Bayesian decision theory*. In later sections there will be reason to use the Bayesian terminology.

of the possible ways the world may be, including when information or choice opportunities will arrive. These ways in which the agent interacts with the world are typically represented in sequential-decision format. Strictly speaking, they should be included in the algebra of propositions of which the agent is aware.

Finally, we note that the sequential-decision format has brought to light some controversies amongst decision theorists concerning how a rational agent should identify and evaluate options.[13] But we can put these controversies aside for the time being, at least, since we are thus far concerned with orthodox agents who maximise expected utility, plan to learn in accordance with the rule of conditionalisation, and moreover expect their plans to be carried through. Under these assumptions, the various approaches to choosing in the sequential-decision context coincide.

2.3 Feedback from the World

As time progresses, the world provides feedback on an agent's roadmap or (sequential-)decision model. Think of this feedback as a learning event that was either planned for or not planned for. In addition, the substance or content of the feedback – the contingency in question – may be either familiar and so yielding an ordinary case of learning (which a rational person responds to by conditionalisation) or else unfamiliar and so yielding awareness growth. So, when the substance of the feedback is unfamiliar, it is a contingency that *was* inaccessible to the agent before she received the feedback.[14]

We should note however that a contingency may be inaccessible – and thus unfamiliar if learned – *only with respect to the specific decision problem*, in which case it is not what we will call 'radically unfamiliar' (and to which we return in a moment). For instance, recall that we assumed that the possibility that the released moth would eat a native plant species was not accessible to our conservationist when reasoning about releasing the moth. So, in that decision problem, the possibility in question was unfamiliar to her. However, this does not mean that the possibility in question was beyond the conservationist's conceptual repertoire; it simply did not occur to her when making this decision.

---

[13] For a summary of the controversies, see Steele (2018).

[14] We mentioned earlier that, strictly speaking, predicted learning events should feature in an agent's proposition space or algebra. It follows that learning events too can be inaccessible to an agent and thus unfamiliar when they arise, as per unplanned learning. In what follows, however, we do not draw attention to awareness growth with respect to the very experience of learning; we simply contrast 'planned and unplanned' learning and refer only to awareness growth with respect to what is learnt.

Table 5 Feedback on our reasoning roadmaps

	familiar content	unfamiliar content
planned	planned conditionalisation	planned awareness growth
unplanned	unplanned conditionalisation	unplanned awareness growth

The observation that learning can be either planned or unplanned, and that the learnt content can be familiar or unfamiliar, suggests that we need to distinguish between four major kinds of feedback from the world, of which Table 5 provides a summary. The more obvious are the two extreme kinds of feedback. The happy case (top left) is where the world effectively vindicates one's model. All learning was planned, and since it concerns familiar contingencies the agent can respond by conditionalising on what was learnt. The less happy case (bottom right) involves learning that was unplanned and moreover concerns unfamiliar contingencies. This is the paradigmatic case of limited awareness and subsequent awareness growth. Finally, there is the middle ground (top right and bottom left).

We will expand on Table 5 in what follows. The point to note from the outset is that awareness growth amounts to gaining recognition of unfamiliar contingencies. While this kind of learning about the world is often unplanned, Table 5 floats the possibility that this need not be so. We can plan for awareness growth. (And as such there may be a case for representing such plans in a sequential-decision model by way of some further symbol, say a diamond.) This issue will be taken up later in the Element.

2.3.1 The Two Extremes

Sometimes the world affirms one's roadmap or model, in the sense that one expects to receive some kind of input from the world, that is, to learn something, and does indeed receive this input. We referred to this as the happy case. Consider our conservationist as described in Figure 1. It may well be that she learns about the drought – whether it is surely happening or surely not happening (and nothing more) – prior to making her choice, just as she predicted. So she learns as planned and the content is familiar.

Of course, part of the reason that she does not perceive other events that her roadmap did not account for may be that she is not looking for these events. Her observations of the world are no doubt 'theory-laden' or rather 'roadmap-laden'. We are not suggesting that, when an agent perceives that the world affirms her roadmap, it has passed some objective test and is shown

Table 6 Conservationist's roadmap after awareness of pest contingencies

	eats; drought; pest	eats; drought; no pest	eats; no drought; pest	...
Release moth	eradicated, pest	eradicated, no pest	eradicated, pest	...
Status quo	thriving	thriving	thriving	...

to be faithful to reality. It is purely the agent's own perspective that we are considering here. What we draw attention to is just that sometimes an agent's roadmap appears to serve her well. The feedback from the world does *not* call for a restructuring of her roadmap but rather affirms it. That need not always be the case, as we will explore now.

What does the less happy case look like? Assume that our conservationist reasons as before, as described by Figure 1. But the feedback she receives from the world is rather different. She learns about the drought, as before, but then a little later she realises there are further contingencies that are pertinent to her choice problem. Some interaction with the world, perhaps simply the sight of a moth landing on a native plant, prompts the realisation that the moth may or may not itself become a pest, eating and destroying native plants.

So in this latter case our conservationist's roadmap turns out not to serve her very well. She has an unplanned learning experience, namely the realisation sparked by the moth landing on the leaf of a native plant. Moreover, this realisation or learning event concerns potential properties of the world that are unfamiliar, in the sense of not being already articulated by her roadmap. It prompts a restructuring of her roadmap, presumably such that her impending decision problem now has the complexity of the third decision problem of the introductory section, reproduced in abbreviated form in Table 6.

We suggest that this sort of roadmap failure and subsequent restructuring is the natural way to think about limited awareness and subsequent awareness growth. That is, it is natural to think that awareness growth is unplanned (although it need not always be, as we shall soon see). As previously mentioned, in the case we described, the newly realised potential properties of the world are in a sense unfamiliar because they are not accounted for in her roadmap, but they are not *radically* unfamiliar. It is not as if our conservationist lacks the concept of either moths or pest species. After all, we stipulated that she cares single-mindedly about eradicating weeds and pests and is contemplating whether to release a moth. Moreover, it is presumably not a giant conceptual leap to imagine moths being pest species, in addition to prickly pears being weeds. Our conservationist simply had not considered, in the context of the choice problem at hand, whether the moth might itself become a pest. Perhaps for very mundane reasons, this simply did not cross her mind, and so was not part of her reasoning, or roadmap, at the time in question.

Later, in Section 4, we will introduce more explicitly cases of limited awareness and subsequent growth that do in fact involve radically unfamiliar properties of the world. The examples concern unusually radical shifts in scientific world view, for instance the shift in the early 1900s to accommodate Einstein's theories concerning the relationship between space and time, amongst other things. These more dramatic examples are useful for highlighting that, at the earlier time, the contingencies in question are truly beyond the agent's grasp of the ways the world might be. We can all appreciate that someone in the early 1900s simply does not have access to the notion that space may be curved, or that someone in the 1970s does not have access to the notion of data sharing over the internet. These are very obvious limitations in awareness. What we draw attention to is that the more ordinary limitations in awareness, like that experienced by our conservationist, may be similarly unavoidable. For whatever reason, sometimes quite ordinary properties of the world are simply inaccessible to an agent at a given point of decision, either due to something as mundane as inattention or lack of imagination or due to more radical limitations in how things are conceptualised.

Since we classify both the ordinary and the more radical unfamiliarity as *unawareness*, we think that a formal model of limited awareness and changes in awareness should accommodate both types. And, indeed, the model we develop in Section 3 is meant to accommodate both types of (changes to) awareness for a given decision situation.

2.3.2 The Middle Ground

We suggested that there are yet two further kinds of feedback that the world may present to an agent, as regards her reasoning roadmap (refer back to Table 5). Our contention is that whether or not the learning experience is planned is orthogonal to whether or not the content of this learning experience is unfamiliar. One can have a planned learning experience of unfamiliar content, giving rise to planned awareness growth. Before we get to this case, however, let us warm up with the other middle-ground case: when the agent learns something familiar but the learning experience was unplanned. This case should strike the reader as very common, even if it is not explicitly acknowledged or discussed in the literature.[15]

To illustrate unplanned learning of the familiar, let us again appeal to the plight of our conservationist. Assume this time that she does not expect to learn anything relevant before making her choice. Her roadmap is simply Table 2, or Figure 1 *without* the circle node representing the learning experience regarding the drought. Now, it may turn out that she does not learn anything; the world affirms her roadmap. Another possibility, however – the one we draw attention to now – is that she unexpectedly learns about the drought; that is, she either learns D or $\neg D$. These are familiar contingencies in that they are represented in her roadmap; for instance, the roadmap in Table 2 includes the contingencies D and $\neg D$. But the learning experience was unplanned, so in this sense her roadmap let her down; for instance, the roadmap in Table 2 doesn't include the learning experience about D and $\neg D$. Presumably, it is easy for her to adapt, however, to her unplanned circumstances. For starters, she can simply proceed to update her credences in the rational way, by adopting her 'old' credences conditional on what she now knows to be true, whether D or $\neg D$. So, the idea is that when it comes to unexpected learning of the *familiar*, the agent does have credences conditional on what she learns (even though the learning was unplanned) which of course implies that what the agent learns was all along in her algebra. But since the learning was *unplanned*, she had not anticipated that she would revise her beliefs in line with these conditional credences in the roadmap she had been using for the decision at hand.

We turn now to the case of planned awareness change: when an agent predicts or plans that she will come to recognise unfamiliar contingencies. We suggest that such occurrences can be represented in a suitably embellished sequential-decision model, even if the contingencies in question cannot be articulated in advance. For instance, our conservationist might anticipate that she will (or at least may) learn something pertinent to her choice problem, in addition to whether or not there is a drought. But she cannot put her finger on what this might be. Her characterisation of the contingency may be more or less abstract. We will not say too much more about this sort of case for now, as the focus of Sections 6 and 7 is what an agent can anticipate and how it affects her current

---

[15] Learning is often discussed in the literature in a way that abstracts from whether or not the learning experience was planned. Sequential-decision models *do* include learning experiences, but, as noted earlier, these models are often treated as objective representations of a temporally extended choice problem rather than a subjective representation that may or may not turn out to be correct.

credences. The answers to these questions are not obvious. The 'hard' position is that limited awareness cannot be well understood by the agent, and it is incoherent to regard her as anticipating awareness growth. As mentioned, we will argue for the 'soft' position that awareness growth can in some sense be anticipated and planned for. But the details matter.

2.4 Concluding Remarks on Section 2

So far we have introduced (un)awareness and changes in awareness in general terms, with reference to the experience of unfamiliar contingencies. We have moreover situated changes in awareness within a general picture of how one's (sequential-)decision model, or roadmap, may be vindicated or else undermined by the feedback one receives from the world. We have not, however, yet said much about how to formally represent limited awareness or how to formally model changes in awareness. That is the main topic of the next section.

3 Modelling (Un)Awareness

3.1 Introduction

We turn now to a closer examination of the forms that limited awareness and changes in awareness might take. We initially, in Section 3.2, introduce a new example which we use to illustrate two intuitively quite different ways in which awareness might grow. Subsequently, in Section 3.3, we again use this example to describe two approaches to modelling limited awareness and growing awareness: first, a model that may seem natural but which we nevertheless reject; next, our own preferred model. Section 3.4 reflects on how our discussion has been informed by others; we briefly survey the diverse treatments of awareness growth in the literature.

3.2 Rent or Buy? Types of Awareness Change

Suppose that you are contemplating buying an apartment and moving out of your rental apartment. The reason is that you have heard that the rent might go up, and you are primarily concerned with whether you will be able to make ends meet. So, you are trying to figure out whether you will be more likely to make ends meet in your current rental apartment or in an apartment that you own. Your decision problem can be represented by Table 7.

Now you realise that an additional important possibility, which you should factor into your decision, is that the owner of the apartment you currently rent might decide to sell, in which case you will, let us assume, find yourself suddenly homeless as the apartment will no longer be available to rent.

Table 7 Rent or buy?

	Available at higher rent	Available at same or lower rent
Rent	Rent & Rent higher	Rent & Rent same or lower
Buy	Buy & Rent higher	Buy & Rent same or lower

Table 8 Expanded rent-or-buy

	Available at higher rent	Available at same or lower rent	Unavailable since sold
Rent	Rent & Rent higher	Rent & Rent same or lower	Rent & Rental apartment sold
Buy	Buy & Rent higher	Buy & Rent same or lower	Buy & Rental apartment sold

This transforms your decision problem[16] into one that can be represented by Table 8.

Before you make your choice, you hear speculations about the central bank planning to raise interests rates. As you realise that this would affect whether you are able to make ends meet after having bought an apartment, you want to factor this possibility into your deliberation too. To simplify the table, let us assume that the interest rate only affects you if you decide to buy. Then the decision problem you are now faced with can be represented by Table 9.

Note that in the shift from the epistemic state represented by Table 7 to the one represented by Table 8, you have *expanded* or *extended* the possibilities you entertain to include the possibility that the apartment is sold. By contrast, in the shift from the epistemic state represented by Table 8 to the one represented by Table 9, you have *refined* some of the possibilities you entertain to accommodate the possibility of a changed interest rate. In this case, your old possibilities are effectively split into more fine-grained ones, allowing for new partitions of the possibility space.

The examples above are special cases of awareness growth: what we might dub *pure* expansion and *pure* refinement respectively. Throughout the Element we will appeal to these sorts of cases. But it is also possible for awareness growth to be a *mixture* of expansion and refinement. For example, one may

---

[16] In keeping with our subjectivist approach, we take an agent's *decision problem* to be (at least) partly defined by the agent's own epistemic state, in particular, what she takes to be her options and the relevant states of the world.

Table 9 Expanded and refined rent-or-buy

	Available at higher rent		Available at same or lower rent		Unavailable since sold	
Rent	Rent & Rent higher		Rent & Rent same or lower		Rent & Rental apartment sold	
Buy	Buy & Rent higher Interest same or lower	Buy & Rent higher Interest higher	Buy & Rent same or lower Interest higher	Buy & Rent same or lower Interest same or lower	Buy & Rental apartment sold Interest higher	Buy & Rental apartment sold Interest same or lower

recognise that the apartment might be sold at the same time as recognising that the interest rate might change (a transition from Table 7 directly to Table 9). Moreover, there may be cases of awareness growth that are even more complicated; for instance, they may involve deleting or retracting possibilities that one had entertained, in addition to expanding and/or refining one's possibilities. We do not explicitly consider any such more complicated cases in this Element. As will be seen, even the simple cases of awareness growth reveal much about how awareness growth can affect a rational agent's credences and desires.

Finally, there are cases of limited awareness and awareness growth that would be more naturally characterised as being about (un)awareness of *options*, rather than being about (un)awareness of states and/or outcomes which has been the focus in this section. We already mentioned such examples in the introductory section. For instance, when discussing the conservationist example we mentioned that from a modern point of view we recognise that the conservationist was unaware of some targeted chemical pest controls. Although limited awareness and growing awareness of options is both theoretically interesting and practically important, we will not in what follows discuss whether this phenomenon calls for special treatment and special norms. As will become apparent, (un)awareness of states and outcomes generates more than enough philosophical problems to occupy us in this Element. And it is plausible, at least, that our model is sufficiently general to accommodate all kinds of limited awareness and awareness growth, whether of states, outcomes or options.

3.3 Modelling Awareness Growth

Tables 7 to 9 above give us a fair idea of what limited awareness and subsequent awareness growth looks like. But in order to get a firm grip on whether an evolving agent like the one described by the transition from Table 7 through to Table 9 is rational, both *at* a time and *over* time, we need to go into a little more detail. Ultimately, we need to describe and assess the agent's evolving credences.

In the introductory section it was noted that a reasoning agent may be depicted as entertaining a set of propositions at the time in question. The set of propositions should form what is known as a *Boolean algebra*; that is, a set of propositions $\mathcal{F}$, that contains both the contradiction and the tautology and is moreover closed under disjunction, negation and conjunction. Moreover, it is assumed that a rational agent has credences in the propositions in $\mathcal{F}$ that can be represented by a probability function P (see Section 1.2.1 for the details).

We want to preserve this way of conceiving a rational agent's epistemic state at any given time but also introduce the possibility of awareness growth. To that end, let us refer to an agent's state of awareness at a given time as her *awareness context*. For any given awareness context, then, the agent entertains a set of propositions – a Boolean algebra – over which she has probabilistic credences. The question is: how does the agent's set of propositions change upon awareness growth? More precisely: what is the relationship between her Boolean algebras, so to speak, from one awareness context to the next? We respond to this question in our own way in what follows, and then, in Section 3.4, we compare our model of awareness growth with those others have proposed.

3.3.1 First Pass: The Catch-All Model

In the philosophical tradition, propositions are typically interpreted as sets of (objective) *possible worlds*, where these worlds are understood to be maximally detailed descriptions of ways the world might be. For instance, the proposition 'Oswald killed Kennedy' is just the set of possible worlds for which that particular proposition is true.[17] The tautologous proposition is thus identified with the set of all possible worlds, while the contradictory proposition is identified with the empty set of possible worlds. Note that it is typically not considered useful to quantify over *all* possible worlds; rather, one quantifies over *the possible worlds relative to $\mathcal{F}$*, which are specific enough just to assign truth values to each of the propositions in $\mathcal{F}$. The set of these worlds amounts to a coarsening of the set of all possible worlds.

This way of conceiving an agent's proposition space, however, does not seem to leave room for awareness growth by expansion. On this model, at any given time, the rational agent supposedly grasps the full set of possible ways the world might be, since she assigns probability one to the tautology – which, in this model, is equivalent to assigning probability one to the full set of (objective) possible worlds. That does not seem to square with transitions, e.g., the shift from Table 7 to Table 9, in which the agent apparently comes to recognise possibilities that are inconsistent with all those she previously entertained – that is, inconsistent with all the propositions in $\mathcal{F}$ – such that there is a genuine enlargement of her (subjective) set of possibilities beyond $\mathcal{F}$. This type of transition cannot be properly modelled as a shift in probability over the adopted set of possible worlds; that is, a set of worlds that allows only for truth value assignments to the propositions in $\mathcal{F}$. It rather involves a revision of $\mathcal{F}$.

---

[17] As the reader may recall, this is not an interpretation we committed ourselves to in Section 1.2.1. (For instance, we did not define the algebra $\mathcal{F}$ with respect to some set of possible worlds; that is, we did not assume that a set of worlds is in any (metaphysical, epistemological or methodological) sense prior to the set of propositions.) But it is an interpretation that is widespread in philosophy.

Table 10 Less aware state with 'catch-all' proposition(s)

	Available at higher rent	**Available at same or lower rent**	**???**
Rent	Rent & Rent higher	Rent & Rent same or lower	Rent & ???
Buy	Buy & Rent higher	Buy & Rent same or lower	Buy & ???

The way to proceed, if one wants to stick with this kind of underlying model, is to engineer the agent's proposition space for any given awareness context so that all instances of awareness growth can be treated, at least formally, as refinements. The thought would be that an agent's algebra includes an abstract *catch-all* proposition signifying 'Other ways the world might be' or 'None of the above'; it represents all those further possible ways the world might be that the agent cannot (yet) articulate. In fact, some argue that an epistemically rational agent would always be open to the possibility that they lack full awareness and thus assign a positive probability to such a catch-all (see, e.g., Chloé de Canson ms). For instance, Table 10 is identical to Table 7 except that it allows for a catch-all proposition that may include 'Other ways in which the landlord affects my living arrangement'.[18]

The idea is that what we earlier called awareness growth by expansion is really a special case of refinement – it is a refinement of the catch-all. In the case that the agent realises that the landlord might sell the apartment (what was formerly the transition from Table 7 to Table 8), the catch-all proposition is effectively divided into 'Rental apartment sold' and, say, 'Other (yet unarticulated) ways the world might be, including how the landlord affects my living arrangement'.

The worry with the catch-all model is that it seems not to be apt as a general characterisation of limited awareness. Nothing that we have said thus far suggests that any agent at all times entertains a catch-all by way of accounting for her inevitably limited awareness. Indeed, this proposal may not even be cogent, since, in order for an agent to make sense of a catch-all, she would presumably need to entertain some universal set of possibilities relative to which the catch-all can be defined as the complement of those possibilities she can properly articulate. But it is hard to see how the agent could have access to this universal set of possibilities (which might in fact not even be a coherent

---

[18] The catch-all will typically be best interpreted as a disjunction, since it must account for *all* other ways that the world might be that are inconsistent with those of which one is already aware. For instance, even for the simple proposition space depicted in Table 10, the catch-all should strictly speaking be 'Other ways in which the landlord affects my living arrangement *or* other housing options that I might choose'.

notion), given that, by assumption, some of these possibilities cannot be articulated. So, it is hard to see how the catch-all could be well-defined for the agent.[19]

By way of response, one might resist these cogency worries. Indeed, far from being incoherent, an agent who accounts for her limited awareness in entertaining a catch-all may be regarded praiseworthy (cf. de Canson ms). Alternatively, one might argue that the cogency worries are moot because a catch-all model need not entail that the agent herself entertains such a proposition. The inclusion of a catch-all proposition, to which the agent implicitly assigns zero probability, may simply be the most elegant way for the modeller to account for an agent's limited awareness. While we are sympathetic to these lines of argument, we do not find them sufficiently convincing (to be elaborated shortly) to continue pursuing a catch-all model of awareness growth.

As for the first: we admit that a certain portrayal of the cogency worry would, as it were, prove too much. The portrayal we have in mind is one that trades on the catch-all being too 'abstract', in that the agent has no idea how to specify the proposition's content. After all, the abstractness of a proposition would seem to be a matter of degree rather than an on/off affair. And one can surely represent an agent as having credences in propositions of varying abstractness without thereby being committed to her being able to precisely articulate what these propositions mean. Moreover, arguably the wiser agents do routinely entertain propositions at the more extreme levels of abstractness – propositions that are intended to capture a wide range of contingencies. Indeed, this point will become important later, in Sections 6 and 7. That said, the *objective catch-all*, as one might refer to it, does not simply capture a wide range of contingencies; it represents *all* other ways the world might be that are inconsistent with those of which the agent is aware. But how can an agent conceptualise *all other ways* the world might be? Therein lies the real cogency problem.

Turning now to the second line of argument: the idea is that the objective catch-all is accessible only to the modeller (or perhaps to the agent herself at a later time), not the agent in question (at that time), who is modelled as implicitly assigning the catch-all zero probability.[20] Refer back to Table 10: the catch-all proposition indicates that the agent has limited awareness. She entertains only two ways in which her landlord may affect her living arrangements in ways relevant to her ends. She fails to see the other possibilities, and thus these

---

[19] We thank Alan Hájek for suggesting this way of putting the problem. He elaborates on worries along these (and other) lines in his unpublished manuscript 'Omega'.

[20] We thank Richard Bradley for this suggestion.

are represented by a catch-all. The agent implicitly assigns the catch-all zero probability. The model in a sense captures a wiser person's perspective and how that person interprets the reasoning of the agent in question. But again, therein lies the problem. For some applications, there may well be special reason to capture a wiser perspective on an agent's limited awareness. But we contend that that is not the case for our application. We are interested in the reasoning of a single agent, who lacks full awareness but is (in other respects) an ideal reasoner, and our aim is to investigate the principles of rationality that govern how her perspective or awareness changes with time. For this purpose, there is no need to keep track of how the agent's awareness looks from some more expansive point of view. Doing so only detracts from the simplicity of the model.

In fact, we think a general model of how an agent should reason in light of her limited and changing state of awareness should not even commit to a *subjective catch-all* (more on which shortly). Such a model would not be a good characterisation of agents who are *unaware of their unawareness*, nor would it be useful for analysing what principles of rationality they should satisfy. We want our general model, which we develop in this section, to leave room for such agents, and thus we do not presume any kind of catch-all. As we explained in the introductory section, our aim is to take as given agents' varying degrees of awareness and ask what principles of rationality these agents should satisfy.

That said, agents may sometimes be *aware of their unawareness* and moreover *anticipate* awareness growth. In fact, in Sections 6 and 7, when we model such agents and consider what rationality principles they should satisfy, we suggest that it is most fruitful to use propositions that may be thought of as *subjective catch-alls*. These types of catch-all are not meant to be interpreted as a metaphysically or logically universal complement to whatever the agent is aware of. Instead, they are meant to represent the possibilities that the agent cannot specify but thinks she might have left out of her reasoning.

3.3.2 Our Preferred Model

The way forward, we suggest, is to divorce the *agent's possibilities* from objective possible worlds. While it is useful to depict an agent's epistemic outlook in terms of atomic possibilities that are the ultimate bearers of probability, these atomic possibilities need not be the objective possible worlds that many think give meaning to propositions. Indeed, a model of an agent's epistemic outlook need not offer an interpretation of propositions. They can simply go uninterpreted. For instance, they can be understood as abstract variables or placeholders that are used to define the 'agent's possibilities'; that is, the possibilities as the agent sees them. That is the approach we shall take. In particular, we define the agent's possibilities as truth functions over these uninterpreted propositions or, more accurately, over the *basic* propositions. We proceed now to spell out our idea more carefully.

We say that an agent's awareness context is defined by a set **X** of *basic* propositions of which she is aware (which we assume to be finite). We take basic propositions to be primitive propositions, representing simple facts about the world, that do not involve any logical connectives. So, for instance, in the awareness context described by Table 7, 'Rent' and 'Rent higher' are basic, while 'Rent & Rent higher' and '¬Rent higher' are not. (Note that, to keep the prose here and in what follows relatively simple, we will use abbreviated expressions to describe aspects of Table 7. So, for instance, 'Available at higher rent' is shortened to 'Rent higher'.)

The basic propositions are not themselves given an interpretation in our model, as previously mentioned; they are simply the primitive facts that the agent is aware of. In other words, any deeper interpretation of these propositions, whether in terms of objective possible worlds or some other kind of structure, is not explicitly modelled here. We do not here take a stance on whether propositions should be identified with sets (see footnote 21). Instead, we simply use them, as abstract and uninterpreted objects, to define the agent's possibilities, to which we turn next.

Let the *possibilities* that the agent is aware of be truth functions, ω_i, that return 'true/false' for each of the basic propositions. Note that we will occasionally use $\omega_1, \omega_2, \ldots, \omega_n$ to denote individual possibilities. The *putative* set of possibilities are all the distinct truth functions that take this form; that is, effectively all the different combinations of truth values for the basic propositions. This is merely the *putative* or *first-pass* set of possibilities, since some will be deemed inconsistent by the agent (to be explained shortly) and thus excluded from the *real* set of possibilities (as recognised by the agent). We may describe the possibilities in terms of conjunctions of the basic propositions for which the ω_i function in question returns 'true'. So, in the awareness context represented by Table 7, the possibility $\{\omega_i(\text{Rent}) = \text{true}, \omega_i(\text{Buy}) = \text{false}, \omega_i(\text{Rent higher}) = \text{true}, \omega_i(\text{Rent same or lower}) = \text{false}\}$ can be described as 'Rent & Rent higher'. From now on, we will use this latter way of describing possibilities.

For the set of basic propositions **X**, let $\mathbf{W_X}$ be the agent's (real) set of possibilities, which is a subset of the *putative* set of possibilities, containing only the possibilities that the agent regards as consistent. A possibility is *consistent*, by the agent's lights, if all its conjuncts *could* be true; that is, if the agent does not take the conjuncts to be mutually inconsistent. What an agent takes to be the set of consistent possibilities will depend on what she regards as partitions of the proposition space (corresponding to properties or categories for which one and only one value can be assumed). For instance, for the agent described by Table 7, one partition of the space is {'Rent', 'Buy'}, these being the candidate values for what we might call the 'action property'; a necessary condition for being a consistent possibility, then, is that the conjuncts include only one of 'Rent', 'Buy'.

So an agent's awareness context X may be just as well defined in terms of her possibility space, W_X. Any given basic proposition X_i can now be associated with a set of possibilities in W_X: the $\omega_i \in W_X$ for which the proposition X_i is true. For simplicity, we refer to this set as $\{X_i\}$.[21,22] We can now also generate a Boolean algebra, $\mathcal{F}_X$, in the usual way: $\neg X_i$ is associated with the set $W_X \setminus \{X_i\}$, $X_i \vee X_j$ is associated with the set $\{X_i\} \cup \{X_j\}$, and $X_i \& X_j$ is associated with the set $\{X_i\} \cap \{X_j\}$. For reasons that will become apparent in the next section, the same proposition can be associated with different sets of possibilities in different awareness contexts. So, more formally, we can think of a proposition as a function from the awareness contexts in which the proposition plays a role to the corresponding sets of possibilities.[23]

The approach described invites some degree of language relativity, for instance, since what counts as a basic proposition (for an agent) may depend on the agent's language. But this is as it should be, we think, given that our starting point is an agent's view of the world – which is of course often shaped by the agent's language – and how that changes as her awareness changes.

For simplicity, we will model only *growth* in awareness over time; our model will not countenance the *shrinking* or *contraction* of awareness over time. Inclusion of the latter possibility would complicate the model and its presentation; moreover, there is a tradition in modelling *rational* belief change to consider only incremental learning (gains in information) rather than forgetting (losses of information).[24] That said, contraction of one's concepts – that is, contraction of the set X – may in some cases not be due to 'forgetting' but rather due to considerations that make it an important aspect of rational

---

[21] We are not here suggesting that the basic propositions are identical to, or defined in terms of, the relevant set of possibilities. After all, the possibilities were themselves constructed from propositions that had some prior meaning. One can retain the traditional notion of propositions being identified with sets of objective possible worlds, as per, e.g., Stalnaker (1984), although this is not explicitly represented in our model. The relation of 'association' that we appeal to here is intended to be weaker than 'identity'.

[22] Strictly speaking, the set in question should be thought of as being indexed to the relevant awareness context. If we wanted to make the index explicit, we could, for instance, write $\{X_i\}_X$. But to simplify the notation, we omit making the index explicit.

[23] To clarify: for awareness contexts where the proposition does not play a role, it is not associated with any set of possibilities.

[24] For notable exceptions, see Titelbaum (2012) and Bradley (2017).

learning. Such contractions may in fact be the correct way of modelling certain kinds of rational learning or realisation, such as when a previous believer in the influence of homeopathy discovers that it has no (non-psychological) influence and so does not number amongst alternative physical treatments for some condition.[25] Due to lack of space, however, we leave further exploration of this phenomenon for future work.

Now let us address the dynamics of awareness. We say that the agent's *awareness grows* when the awareness context shifts from $\mathbf{X}$ to $\mathbf{X}^+ = \mathbf{X} \cup \mathbf{X_j}$. Note that by the assumptions we made earlier, when the awareness context shifts from $\mathbf{X}$ to $\mathbf{X}^+$ there is a corresponding shift from $\mathbf{W_X}$ to $\mathbf{W_{X+}}$ and from $\mathcal{F}_X$ to $\mathcal{F}_{X+}$. Strictly speaking, $\mathbf{W_X}$ and $\mathbf{W_{X+}}$ do not have any possibilities in common; after all, the possibilities in each are truth functions that have a different number of propositions in their domain. If, however, we allow that the possibilities may be described in terms of the proposition that they are each associated with – the conjunction of all basic propositions for which the function in question returns 'true' – then $\mathbf{W_X}$ and $\mathbf{W_{X+}}$ may in certain cases (as we will see shortly) have possibilities in common.

Now we can characterise the difference between awareness growth by expansion and awareness growth by refinement. Let us measure the *length* of a possibility by the number of propositions for which the function in question returns 'true'. (Recall that we assume that the set of basic propositions is finite.) We say that the awareness growth was (purely) due to *expansion* if the *number* of possibilities in $\mathbf{W_{X+}}$ is greater than in $\mathbf{W_X}$, without any possibilities becoming *longer* in the sense given. In contrast, we say that the awareness growth was (purely) due to *refinement* if the *number* of possibilities in $\mathbf{W_{X+}}$ is greater than in $\mathbf{W_X}$, and moreover, at least some possibilities in $\mathbf{W_{X+}}$ are *longer* (in the sense just described) than the possibilities in $\mathbf{W_X}$. Moreover, in the case of pure expansion there are some possibilities common to $\mathbf{W_X}$ and $\mathbf{W_{X+}}$, while, in the case of pure refinement, there are no possibilities common to $\mathbf{W_X}$ and $\mathbf{W_{X+}}$.

Return again to our rent-or-buy example, and suppose now that in the least-aware context (Table 7), the only possibilities that the agent of interest is aware of and considers consistent can be characterised as: 'Rent & Rent same or lower', 'Rent & Rent higher', 'Buy & Rent same or lower', 'Buy & Rent higher'. In other words, she regards any possibility that involves 'Rent & Buy', and likewise 'Rent same or lower & Rent higher', inconsistent. Now, when awareness grows due to an expansion, e.g., when the agent becomes aware of the possibility that the owner sells the apartment – which the agent takes to

---

[25] We thank Sven Ove Hansson for this particular example.

be inconsistent with the owner keeping the rent the same and also inconsistent with the owner increasing the rent – the possibilities do not become longer. Instead, we simply add 'Rent & Rental apartment sold', 'Buy & Rental apartment sold' to the original four possibilities. This is represented by the shift from the awareness context represented by Table 7 to the one represented by Table 8.

In contrast, when awareness grows due to refinement of the possible interest rate – as represented by the shift from the awareness context represented by Table 8 to the one represented by Table 9 – some of the new possibilities are longer; for instance, 'Buy & Rent same or lower & Interest higher' compared to 'Buy & Rent same or lower'. The number of possibilities also grows, since, e.g., 'Buy & Rent same or lower' becomes 'Buy & Rent same or lower & Interest higher', 'Buy & Rent same or lower & Interest same or lower'.

3.4 Awareness Growth in the Literature

We noted in Section 1 that there is already a small but significant literature on (un)awareness and awareness growth, dating back thirty years or so. A select group of economists and computer scientists have introduced this phenomenon to their various abstract models of reasoning and knowledge (and indeed, as noted earlier, the term *(un)awareness* comes from this literature).

For the most part, at least until very recently, philosophers have tended to consider awareness growth in relation to very specific questions in the philosophy of science concerning theory change. In the last few years, however, occasional philosophers have turned their attention to awareness growth more generally. Richard Bradley's (2017) recent work for instance pursues the more general challenge that (un)awareness poses for decision theory. As will become apparent in the next section, we criticise aspects of Bradley's general approach to unawareness. Others have also weighed in on this debate, including Mahtani (2020) and de Canson (ms) (whose work we mention in this and the next section). Yet others have applied Bradley's general framework to particular philosophical debates, for instance Aron Vallinder (2018), who examines the problems that changing awareness poses for moral epistemology, and Joe Roussos (2020), who discusses the relationship between expert testimony and awareness growth.

Our treatment of (un)awareness draws on various aspects of earlier approaches. Like Bradley (2017) and economists such as Karni and Vierø (2013, 2015), our aim is to characterise a (single) rational agent's reasoning in the face of limited awareness and subsequent awareness growth. Note that this need not be the aim: the pioneering work on (un)awareness by Fagin and

Halpern (1987), for instance, and the work of many economists since, is more about characterising the relative state of (un)awareness of several agents, e.g., in the context of a game.

Our approach is also similar to Bradley's and Karni and Vierø's in that although we examine both practical and epistemic reasoning, we take the decision-perspective to ground our analysis. Ultimately, our aim, just like theirs, is to figure out how a *decision-maker* ought to reason in light of her limited and changing state of awareness. In this our approach differs from the early work of philosophers of science concerning the challenge that growing awareness poses for the traditional probabilist model of belief and confirmation under the guise of 'the problem of new theories' (Earman 1992) or 'the problem of old evidence' (Glymour 1980). (Later in this section we mention another difference between our model and theirs.)

As is typical of philosophers, we prefer the generality of a decision framework inspired by Jeffrey (1965), rather than the Anscombe and Aumann (1963) framework (based on that of Savage 1954) employed by Karni and Vierø. In this respect, our approach is close to Bradley's. What is distinctive about the Jeffrey framework is that all events are treated as propositions towards which an agent has both an epistemic and a desire attitude. That is, no distinction is made between the objects of actions, desires and beliefs.[26] But our model has a further subjectivity that is more in keeping with the the approach of some economists and computer scientists: we treat the relevant set of possibilities as dependent on an agent's state of awareness, in that the set changes as awareness grows.

Philosophers have not embraced this kind of 'subjective state space' (as Schipper 2015 puts it, in an important review of economists' and computer scientists' work on (un)awareness). For the reasons discussed in Section 3.3.1, the natural move when working with objective possible worlds is to introduce a catch-all proposition standing for 'the other ways the world might be', where this catch-all need not be something the agent herself entertains. That is the approach taken by various philosophers seeking to model growing awareness about new scientific theories, in the context of the aforementioned problems of new theories and old evidence, including Shimony (1970), Earman (1992), and recently Wenmackers and Romeijn (2016). Bradley's (2017) model does not appeal to a catch-all per se, but the agent's point of view is described in contrast to the modeller's point of view, such that the former can be seen as presupposing some proposition to be true and its negation false.

---

[26] See e.g. Joyce (1999) for a discussion of how to translate between Jeffrey's framework and Savage's.

While we have some sympathy with these catch-all approaches, we question the cogency and/or the usefulness of including an (objective) catch-all in a model of the reasoning of agents with limited awareness (recall our arguments in Section 3.3.1). However, as previously explained, we acknowledge that – at least to some extent – the choice between the model we have proposed and these earlier catch-all models is like any other modelling choice in that it should be judged relative to the purpose to which the model is put. For some purposes, the (objective) catch-all model, appealing to the perspective of a wiser party or modeller, is indeed very suitable.

Other philosophers do not appeal to catch-alls in accommodating (un)awareness and awareness growth but rather abstract propositions that stand in for classes of contingencies whose members are yet to be better articulated. As a result, however, these models are highly constrained in terms of the (un)awareness and awareness growth that is permitted. For instance, Maher (1995) assumes that the agent's algebra contains variable propositions for *each* of the yet-to-be-formulated theories, and he moreover assumes that the agent assigns a (non-zero) probability to each such proposition. Henderson et al. (2010) propose something similar, although with the added sophistication that the propositions in the agent's algebra form a hierarchy that remains fixed throughout the investigation; that is, it remains unchanged even when the agent becomes aware of new theories that effectively fill in this hierarchy.

It is also worth mentioning the proposal of Zabell (1992) in this context. Zabell extends statistical inference to cases where previously unsuspected phenomena of a given kind may occur (such as in the so-called *sampling of the species* problem where an unrecognised species may be sampled). As in Henderson et al., the probability function is defined over a set of hypotheses that are sufficiently abstract to accommodate all the possible phenomena of the given kind, whatever they turn out to be. Moreover, by construction, the probability function does not depend on how exactly the abstract hypotheses are instantiated.

Unlike Zabell, Maher and Henderson et al., we do not assume that the agents we consider always assign a positive probability to some yet-to-be formulated theory or possibility – let alone *all* such theories or possibilities. Instead, we will argue, in Sections 6 and 7, that the appeal to abstract propositions that are placeholders for yet-to-be-fully-articulated contingencies is an apt way to model *only* the special case whereby an agent *anticipates* her own awareness growth. Note that economists too have appealed to 'catch-alls' of this sort. For instance, Grant and Quiggin (2013a, 2013b) incorporate what we call a *subjective catch-all* in their model; it is assigned a probability based on the agent's past experience of limited awareness. In one of their more recent papers,

Karni and Vierø (2017) introduce a subjective catch-all consequence to allow for the agent anticipating her awareness growth, which they also call 'awareness of unawareness'. This psychological phenomenon is also sometimes dubbed 'conscious unawareness' (Walker and Dietz 2011), or 'introspective unawareness' (Piermont 2017). We suggest that these models do not incorporate what is strictly speaking a catch-all, i.e., a proposition standing for *all* other possibilities of which the agent is unaware (as perceived by a wiser party or modeller), but rather appeal to an abstract proposition standing for a broad class of contingencies that the agent believes she may 'fill in' later. The same is true of the subjective catch-all model we introduce in Section 6.

An agent need not always anticipate her own awareness growth, however, such that it can be treated formally as a refinement. We insist that there can be genuine cases of awareness growth by expansion. That is why the notion of a subjective and changing possibility space is important. In this respect, our model draws on the work of economists Heifetz et al. (2006, 2008). In particular, and as previously explained, we draw on the idea that an agent's possibility space is constructed from combinations of the basic propositions of which she is aware, which may change over time. Heifetz et al. (2006: 80) model progressive awareness change in terms of 'a complete lattice of disjoint spaces'. Similar to our characterisation of propositions earlier in this section, Heifetz et al. also refer to a 'surjective projection' from more aware to less aware spaces; i.e., every atomic possibility in the more aware state maps onto a single atomic possibility in the less aware state, whereas some possibilities in the less aware state map onto more than one atomic possibility in the more aware state.

The preoccupations of Heifetz et al. – and indeed the finer details of their model – differs, however, from ours. As per Fagin and Halpern (1987), they focus on (un)awareness in the context of interactions between multiple agents, or games.[27] For them, the question of how to characterise an agent's growing awareness is therefore only the starting point. There are further questions about how knowledge interacts with (un)awareness. What they are ultimately interested in is how rational agents' respective (un)awareness and their knowledge of others' (un)awareness affects the actions that the agents take in some interactive decision problem (i.e., in a game).

3.5 Concluding Remarks on Section 3

The main aim of this section, as the reader will recall, was to develop a model of reasoning and choice that is more subjective than even traditional subjective

---

[27] As discussed, this is not, however, true of *all* economists (nor of all computer scientists) (see, e.g., Grant and Quiggin 2013a; Walker and Dietz 2011; Karni and Vierø 2013, 2015; Piermont 2017).

expected utility theory, in that our model makes room for the possibility that agents differ in what they are aware of and the extent to which they are aware. Although we have now developed such a model, and explained how it can be used to characterise intuitively different ways in which awareness can grow, we have not yet considered how an agent should change her credences when awareness grows. That will be the topic of the next two sections: 4 and 5. Moreover, while we have suggested that agents sometimes can, and arguably should, anticipate that their awareness will grow, we have not examined how they should factor this anticipation into their reasoning. That will be the topic of Sections 6 and 7.

4 Responding to Awareness Growth

4.1 Introduction

So we have introduced awareness growth informally and formally. How should an agent respond to such growth? That is what we turn to now. We consider what sort of impact awareness growth has on the credences (i.e., degrees of belief) of a rational agent.

We know that awareness growth is different from the type of learning that philosophers and decision theorists typically consider. The typical case is where the agent comes to have a new credence in propositions representing familiar contingencies. (Recall our characterisations of different types of learning in Section 2, Table 5.) With awareness growth we are talking about coming to entertain propositions that represent previously unfamiliar contingencies. This is not a learning experience that can be characterised in the usual way: as a constraint on the agent's probability function over a given possibility space. It rather involves a revision of this very possibility space.

This section and the next investigate the extent of the analogy between awareness growth and traditional kinds of learning. One way to put the question is as follows: what is the parallel between norms for belief or credence revision under awareness growth and regular so-called *Bayesian* norms for belief revision?[28] (Recall from Section 2, in particular footnote 12, that the probabilist model of belief, together with the norm of belief revision known as *conditionalisation*, is collectively referred to as the *Bayesian* model of belief, or simply as *Bayesianism*.[29])

---

[28] Throughout this Element, we focus on *degrees of* belief, rather than on outright (all-or-nothing) belief. Thus, when we talk about 'belief revision' we mean revision of degrees of belief, that is, revision of one's credences.

[29] Note that *Bayesianism* is often taken to incorporate the expected utility principle as well.

The standard answer in the literature – at least to the extent that there can be said to be a 'standard' answer when it comes to awareness growth – is that there is a strong resemblance between awareness growth and ordinary learning and similarly between the norm for awareness growth and Bayesian conditionalisation. In the case of awareness growth, the norm has become known as *Reverse Bayesianism*, a term coined by economists Karni and Vierø, who have devised influential decision-theoretic arguments in favour of the norm (2013, 2015). Informally, Reverse Bayesianism states that when a person becomes aware of new contingencies, she should update her credences 'in such a way that likelihood [probability] ratios of events in the original [epistemic] state space remain intact' (2013: 2801). Insofar as they have addressed belief revision under awareness growth, philosophers too (notably, Wenmackers and Romeijn 2016; Bradley 2017) have endorsed what is effectively Reverse Bayesianism.

In this section we investigate the Reverse Bayesianism response to the question posed above. We initially, in Sections 4.2 and 4.3, explain why Reverse Bayesianism has a strong resemblance to regular Bayesian belief revision. We go on, however, in Section 4.4, to pose putative counterexamples to Reverse Bayesianism. Our diagnosis of these cases leads us to conclude that they are indeed genuine counterexamples to Reverse Bayesianism, which thus cannot be a general requirement of rationality.

4.2 Traditional Bayesianism

Let us first lay out in detail the traditional Bayesian account of how credences should be revised in response to learning.[30] We note upfront that the finer details of Bayesian norms of belief revision are controversial. For instance, there is disagreement about whether the norms govern the *actual* transitions in an agent's credences through time, or merely an agent's *plans*, at some given time, for revising her credences in response to learning new things. For many purposes, such finer interpretative questions arguably do not matter. We hope that interpretative issues do not matter for our purposes either, but that is not something we can take for granted. We seek belief revision norms for awareness growth that are the counterpart of norms for ordinary learning. Our response will be more compelling to the extent that it does not depend on a specific and controversial interpretation of these norms for ordinary learning.

---

[30] A very brief account of Bayesian learning was given in Section 2, specifically in Section 2.2.

4.2.1 The Rule of Conditionalisation

We will get to the nuances shortly. Let us first present the basics. The rule of conditionalisation – one of the core theses of Bayesian epistemology – states that for any proposition B, the agent's degrees of belief or credence in B, after learning A (and nothing stronger), should equal her (prior) conditional credence in B given A, i.e., $P(B \mid A)$, which, according to the standard definition of conditional probabilities, equates to $P(A\&B)/P(A)$ whenever $P(A) > 0$. (It is typically assumed that the agent would never learn something to which she had assigned zero probability.[31]) More formally, let P_A represent our agent's credences after she has learned A. Then the rule of conditionalisation states that:

Conditionalisation. *For any* $A, B \in \mathcal{F}$ *and according to any rational agent:*

$P_A(B) = P(B \mid A)$ *assuming that* $P(A) > 0$.

Now, just like we used P_A to denote our agent's credences after she has learned A, we can use it to denote her *conditional* credences after she has learned A. Thus, $P_A(B \mid C)$, for instance, denotes her conditional credence in B given C after she has learned A. Then, given the standard definition of conditional probabilities stated above, conditionalisation is logically equivalent to the conjunction of the following two principles:

Certainty. $P_A(A) = 1$.

Rigidity. $P_A(B \mid A) = P(B \mid A)$ *assuming that* $P(A) > 0$.

Informally, Certainty says that the agent is certain of whatever she has learned. Rigidity, on the other hand, says that whatever proposition the agent may learn, her degrees of belief conditional on this proposition are rigid, or unchanged by the learning experience. These two principles thus reflect a neat division between those beliefs directly affected by the learning experience (described by Certainty) and those beliefs that are not affected by the learning experience and are thus unchanged (as per Rigidity). In this way, the rule can be described as a 'conservative-change' maxim: 'hold fixed the relationships between any beliefs that are not directly affected by the learning experience'.

It has been well noted that the Certainty condition does not encompass all kinds of learning. For starters, it does not fit well with an intuitive notion of

---

[31] We leave it as an open question whether or not the relevant conditional probabilities can otherwise be defined for cases where the proposition conditioned on is assigned zero probability (see Hájek 2003 for discussion). In general, there is a question which we do not address in this Element of how an agent should revise her credences if she were to learn that a proposition which she had assigned zero probability is in fact true.

learning according to which one could take oneself to have learned something without having become certain of some proposition. Fortunately, the Bayesian framework can be straightforwardly extended to learning experiences where an agent does not learn anything with certainty, without giving up Rigidity, as Richard Jeffrey (1965) proposed.[32]

4.2.2 Interpreting the Norm

We said that conditionalisation articulates conservative belief change. One might defend such a rule on the basis that an agent should adjust her credences in response to learning only to the extent that the learning is revelatory. Any further change would be arbitrary. For instance, when an agent learns a proposition A with certainty, the only change that this licences is, first, a credence of 1 in A, and, second, the corresponding credence change in other propositions brought about by conditionalising on A. All other credences should stay the same. In particular, all conditional credences should stay the same (as per Rigidity), and so should the credence in any proposition B that the agent takes to be independent of A in the sense that conditionalising on A does not change her credence in B.

One might wonder why an austere approach to learning is rationally required, and, if so, whether an austere approach amounts to conditionalisation. But let us put this big substantive question to the side. We focus rather on the interpretative question introduced earlier: is the relevant norm about transition or planning? That is, are we talking about whether an agent should *actually* or rather merely *plan to* change her credences according to the rule of conditionalisation?[33] The hope would be that it does not matter: either interpretation of Bayesian conditionalisation will have an analogue in the case of awareness growth.

Indeed, defenders of the Bayesian model (so-called *Bayesians*) have taken different sides on questions of interpretation. The standard reading of conditionalisation is arguably the transition reading: that it is a genuinely diachronic norm governing the relationship between an agent's credences at different times, before and after learning. The planning interpretation is more attractive,

---

[32] The maxim has also been extended to cases where the information one gains affects one's conditional probabilities (in the form of the rule known as *Adams conditionalisation*) (Bradley 2005).

[33] Others have suggested that there is no direct norm governing credence change at all. Rather, conditionalisation is a mere consequence of rational agents having credences that fit the evidence at any given time (Hedden 2015). But this picture involves substantial evidential constraints on credence and lies well outside the subjectivist picture we are pursuing here.

however, to those who are uncomfortable with the idea of genuinely diachronic norms of rationality and who doubt that such norms can be defended.

The problem, when it comes to remaining ecumenical about the interpretation of belief revision norms, is that any planning norm does not seem to have an obvious analogue in the context of awareness growth. There is no sense in which an agent can plan for a *particular* awareness growth, as that would contradict the very nature of the phenomenon. That said, one might suppose that an agent can plan for some generically described awareness growth. Indeed, we explore such plans in detail in Sections 6 and 7, under the auspices of 'anticipated awareness growth'. Accordingly, one might regard the norm we seek, the analogue of conditionalisation, to be a constraint on one's planned credence change in the event of some generically described awareness growth. While we do return to this idea in Section 7.4, for now it is worth noting that there are certain limitations to the kinds of awareness growth that one can anticipate. For instance, one cannot anticipate what is truly awareness growth by expansion. It seems that, insofar as we are looking for a parallel norm for *all kinds of* awareness growth, it is more straightforward to have in mind the actual-transition version of Bayesian conditionalisation.

It is however worth asking: is there yet another way to interpret norms of belief change in the standard setting that can also be extended to awareness growth? Let us suggest one that accords with the conservative nature of conditionalisation. The main idea is that conditionalisation is not a substantive norm at all but rather a mere description of what it means for credences to be stable upon learning. That is, if an agent who learns some proposition for sure changes her credences in accordance with conditionalisation then her credences are stable throughout the experience. This is the most deflationary account of the significance of 'conservative belief change'.

On this reading, an agent is not irrational in any sense if her actual changes in credence do not accord with conditionalisation. It is just that her credences are not stable, or do not change in a stable way, at the time in question. (One might, in addition, be interested in an agent's overall epistemic stability. For instance, one might regard a person to be an epistemically stable agent even though she *occasionally* fails to conditionalise, as long as she doesn't do so too often.) Similarly, in the case of planning, it is not irrational for an agent to predict that she will not revise her credences in accordance with conditionalisation. There may nonetheless be something regrettable about this kind of self-knowledge since there may be costs associated with instability in one's credences.

We have introduced this latter interpretation of Bayesian conditionalisation to allow for at least two ways to think about what a 'norm' of belief change for growing awareness amounts to. That is to say that one can read the quest of this section in at least two ways. One option is that we seek a genuinely diachronic norm for belief change under growing awareness. That was the presumption in our introductory remarks. We asked: how should an agent change her credences in response to a given growth in awareness? Another option, for those sceptical of genuinely diachronic norms (and who moreover doubt the prospects for the relevant 'planning' norm), is that we seek a mere description of, or criteria for, what it means for credences to be stable through time, having experienced awareness growth. We will proceed using language that is more fitting to the former interpretation, but everything we say can rather be read along the lines of the latter.

4.3 Reverse Bayesianism

The 'standard' response to the question of how one should change one's credences in response to awareness growth, is, as mentioned, known as Reverse Bayesianism. We note upfront a fixed point in our discussion of Reverse Bayesianism and potential rivals. We take *probabilism* to be non-negotiable: we are assuming that a rational agent's credences *in any given awareness context* must satisfy the probability axioms. Recall from Section 3.3 that an *awareness context* is defined by the set of *basic propositions* of which the agent is aware, from which a Boolean algebra can be generated.[34] A brief comment is in order here: the tautology, which has probability one according to the probability axioms, must be interpreted such that it depends on the awareness context: it is associated with the set of all (epistemic) possibilities in that context, which corresponds, for instance, to $A \vee \neg A$, for any A in the context.

The question is whether and how an agent's subjective probability function for one awareness context constrains or relates to her subjective probability function once she has experienced a growth in awareness. Let's consider a variant of the rent-or-buy example from the last section. This new example involves a slightly more complicated set of contingencies. In your least aware epistemic situation, represented by Table 11, you perceive the outcome of your decision as to whether to continue to rent or instead buy an apartment to depend not just on how the rental price will change but also on whether the new neighbours would turn out to be noisy or quiet. Note that in order to make these two sorts of properties of the world vivid, Table 11 suppresses the possible actions of 'Rent' and 'Buy'; the table is rather a two-dimensional representation of the state space.

---

[34] In Section 3.3 we specified that we understand *basic propositions* to be primitive propositions (representing simple facts about the world) that do not involve any logical connectives.

Table 11 State space for rent-or-buy

	Available at higher rent	**Available at same or lower rent**
Neighbours noisy	Neighbours noisy & Rent higher	Neighbours noisy & Rent same or lower
Neighbours quiet	Neighbours quiet & Rent higher	Neighbours quiet & Rent same or lower

Now suppose that having found yourself in the least-aware epistemic situation, you become aware that your rental apartment might be sold, as per Table 12. How should you revise your credences in the various other propositions in light of this expansion? Or, instead, suppose that in the situation represented by Table 12, you realise that changes to the interest rate affects the outcome of buying an apartment, as per Table 13. How should this refinement affect your credences in other propositions?

Traditional Bayesianism is silent on these two questions. As we have seen, this is not a type of learning experience that the traditional Bayesian framework incorporates. But recently, Karni and Vierø have defended a unified answer to these two questions (at least for the particular kind of decision problem and awareness growth that they represent) in the form of a principle that they call 'Reverse Bayesianism'.

Let us state Reverse Bayesianism as if it were a general principle transcending the particular type of decision model formulated by Karni and Vierø (more on which later in this section). We use P and P^+ respectively to represent the probabilistic degrees of belief of the agent before and after awareness grows. As per the notation in Section 3.3, $\mathbf{X}$ ($\mathbf{X}^+$) is the set of basic propositions of which the agent is aware before (after) awareness grows. Reverse Bayesianism holds that the ratio between the probabilities of any two inconsistent[35]

---

[35] The reader might wonder why the propositions are required to be inconsistent. (We thank a referee for encouraging us to clarify this.) The reason is that otherwise Reverse Bayesianism, stated in a proposition framework like that adopted in this Element, will not hold for awareness growth by expansion unless the agent in question only recognises one way of partitioning her algebra (which of course few agents ever do). For instance, note that in the awareness context represented by Table 11, the agent recognises two partitions; one in terms of whether the new neighbour is noisy or not, another in terms of how the landlord might affect the result of not buying a new apartment. Now, in the shift from Table 11 to Table 12, the agent becomes aware of a possibility that is not accounted for in the latter partition; so, if the newly discovered possibility gets a positive probability, the probabilities of the possibilities already accounted for in that partition ('Available at higher rent' and 'Available at same or lower rent') must shrink. In contrast, the probabilities of the possibilities in the first partition ('Neighbour noisy' vs. 'Neighbour quiet') are unaffected. So, without the restriction to inconsistent propositions, Reverse Bayesianism would fail in this case – and, more generally, in any case of expansion where the agent recognises more than one partition.

Table 12 Expanded state space for rent-or-buy

	Available at higher rent	Available at same or lower rent	Rental apartment sold
Neighbours noisy	Neighbours noisy & Rent higher	Neighbours noisy & Rent same or lower	Neighbours noisy & Rental apartment sold
Neighbours quiet	Neighbours quiet & Rent higher	Neighbours quiet & Rent same or lower	Neighbours quiet & Rental apartment sold

Table 13 Expanded and refined state space for rent-or-buy

	Available at higher rent	Available at same or lower rent	Rental apartment sold
Neighbours noisy	Neighbours noisy & Rent higher & Interest higher Neighbours noisy & Rent higher & Interest same or lower	Neighbours noisy & Rent same or lower & Interest higher Neighbours noisy & Rent same or lower & Interest same or lower	Neighbours noisy & Rental apartment sold & Interest higher Neighbours noisy & Rental apartment sold & Interest same or lower
Neighbours quiet	Neighbours quiet & Rent higher & Interest higher Neighbours quiet & Rent higher & Interest same or lower	Neighbours quiet & Rent same or lower & Interest higher Neighbours quiet & Rent same or lower & Interest same or lower	Neighbours quiet & Rental apartment sold & Interest higher Neighbours quiet & Rental apartment sold & Interest same or lower

basic propositions in the old epistemic state (that each had positive probability) should not change when awareness grows. More formally:

Reverse Bayesianism. *For any $A, B \in \mathbf{X}$ (where $P(A\&B) = 0$, $P(A) > 0$ and $P(B) > 0$) and according to any rational agent:*

$$\frac{P(A)}{P(B)} = \frac{P^+(A)}{P^+(B)}.$$

Versions of this principle have more recently been endorsed by Wenmackers and Romeijn (2016) and Bradley (2017).

Consider what Reverse Bayesianism requires in the rent-or-buy variant. Suppose you find 'Neighbours noisy' to be twice as likely as 'Neighbours quiet' before realising that the apartment could be sold. Then after this realisation, you should still find 'Neighbours noisy' to be twice as likely as 'Neighbours quiet'. Similarly, after you realise that the central bank might change the interest rate, you should still find 'Neighbours noisy' to be twice as likely as 'Neighbours quiet'. On the face of it, these implications of Reverse Bayesianism might seem intuitive. For why should the prospect of the the apartment being sold, say, change one's relative credence in whether the new neighbours will be noisy vs. quiet?

One might surmise that Reverse Bayesianism is compelling because it precisely captures conservative belief change for the learning experience in question – awareness growth. Indeed, its defenders take it to be the consequence of something akin to the Rigidity condition for this kind of learning experience. Bradley, for instance, says as much:

> Within the Bayesian framework, conservation of the agent's relational beliefs is ensured by the rigidity of her conditional probabilities. So we can conclude that conservative belief change [when faced with growing awareness] requires [that] the agent's new conditional probabilities, given the old domain, for any members of the old domain should equal her old unconditional probabilities for these members. (2017: 258)

Wenmackers and Romeijn similarly suggest that the conservation of 'probability ratios among the old hypotheses' follows from the relevant conditional probabilities remaining constant:

> In analogy with Bayes' rule, one natural conservativity constraint is that the new [i.e., more aware] probability distribution must respect the old [i.e., less aware] distribution on the preexisting parts of the algebra [i.e., on the distributions' shared domain]. (2016: 1235)

Karni and Vierø also appeal to the constancy of conditional attitudes by way of defending Reverse Bayesianism. In the behaviourist economics tradition, they appeal to constraints on preferences and only indirectly on beliefs:

> as the decision-maker's awareness of consequences grows and his state space expands, his preference relation conditional on the prior state space remains unchanged. (2013: 2801)

These defences of Reverse Bayesianism are arguably sound given the models of awareness growth to which they pertain. As mentioned earlier, however, these models place limitations on the kinds of awareness growth and/or the beliefs that are subject to the Reverse Bayesianism rule. Karni and Vierø, for instance, employ an Anscombe and Aumann (1963) framework, which (like Savage's 1954 framework, on which it is based) consists of acts, maximally specific consequences and states amounting to act-consequence pairs. For Karni and Vierø, in cases of awareness growth by expansion, the agent ultimately comes to be aware of states that are by their very nature inconsistent with the states that define her old awareness context.

Philosophers tend to prefer a more general propositional framework inspired by Jeffrey (1965) but nonetheless introduce similar restrictions to Karni and Vierø in their discussions of belief change under growing awareness. Wenmackers and Romeijn (2016) focus on changes to sets of scientific theories that are assumed to be mutually inconsistent. Bradley's interests are more general, but he too, in his endorsement of Reverse Bayesianism for awareness growth by *expansion*, at least, focuses on propositions that are inconsistent with those the agent comes to be aware of:

> the key to conservative attitude change in cases where we become aware of prospects *that are inconsistent with those that we previously took into consideration* is that we should extend our relational attitudes to the new set in such a way as to conserve all prior relational beliefs (2017: 257, emphasis added)

Moreover, we hold that Bradley implicitly assumes only 'vanilla' kinds of awareness growth by refinement, as our discussion in Section 4.4 will reveal.

We allow that Reverse Bayesianism may be defensible in the limited setting that these authors consider. But the question remains as to whether this learning rule is defensible in a more general setting (as Bradley, at least, suggests). In the next section, we show that the answer to this question is negative: we offer some informal counterexamples to Reverse Bayesianism.

4.4 Counterexamples to Reverse Bayesianism

It is not hard to see that Reverse Bayesianism cannot generally be true once we move beyond the constrained models of its defenders. That is, one can devise examples where Reverse Bayesianism is violated without irrationality on behalf of the agent in question. All we need are examples where awareness grows since an agent becomes aware of a proposition that she takes to be evidentially relevant, intuitively speaking, to the comparison of propositions of which she was already aware. For in that case, the ratio between probabilities of propositions of which the agent was already aware will not stay the same; one proposition will become more probable compared to the other, just like in ordinary cases where one learns evidence relevant to the comparison of hypotheses.

In fact, the history of science is full of examples that undermine Reverse Bayesianism, for this very reason. Here is a particularly prominent such example:

Example 1. Nineteenth-century physicists were unaware of the Special Theory of Relativity (STR). That is, not only did they not take the theory to be true; they had not even entertained the theory. We can suppose, however, that they had entertained various propositions for which the theory was regarded evidentially relevant, once Einstein brought the theory to their attention. In particular, they did (rightly) take the theory to be evidentially relevant to various propositions about the speed of light, such as whether the speed of light would always be measured at 300,000 km/s independently of how fast the investigator is moving or whether the measured speed would differ depending on how fast the investigator is moving. But then the awareness and subsequent acceptance of the STR changed their relative confidence in such propositions.

Not all examples where Reverse Bayesianism fails come from the history of science. Here is a more mundane, or everyday, example:

Example 2. Suppose you happen to see your partner enter your best friend's house on an evening when your partner had told you she would have to work late. At that point, you become convinced that your partner and best friend are having an affair, as opposed to their being warm friends or mere acquaintances. You discuss your suspicion with another friend of yours, who points out that perhaps they were meeting to plan a surprise party to celebrate your upcoming birthday – a possibility that you had not even entertained. Becoming aware of this possible explanation for your partner's behaviour makes you doubt that she is having an affair with your friend, relative, for instance, to their being warm friends.

There is an important difference between the two examples. In the first example, the awareness growth consists in the recognition of a completely new idea that the scientists had never heard or thought about before. In the second example, by contrast, the agent has, we can assume, heard of surprise parties before; it is just that in the situation and the moment in which he/she finds him/herself, the possibility is not part of his/her awareness. As we however pointed out in the introductory section, our interest is the role that an agent's epistemic state plays in her deliberation about how to act. And since 'unawareness' due to never having heard about an idea generally plays the same role in deliberation as 'unawareness' merely in the moment of deliberation, we treat these two types of epistemic limitations in the same way.

A defender of Reverse Bayesianism might argue that these two examples do not undermine their thesis, since, for instance, the proposition picked out by the sentence 'the speed of light will always be measured at 300,000 km/s independently of how fast the investigator is moving' is different before and after the speaker becomes aware of the Special Theory of Relativity. (Similarly, the proposition picked out by the sentence 'my partner and best friend are having an affair' is different before and after the speaker realises that their partner and best friend might be organising a surprise party.) That is, it is not just that the propositions in question are *understood differently*, given a change in the underlying possibility space (as per our own approach, detailed in Section 3.3); rather what appear to be the same propositions across awareness contexts are in fact entirely different propositions. For instance, the physics case might be spelled out as follows: despite appearances, the agent's growth in awareness is not simply an expansion of the 'fundamental physical theory' partition to include the STR; there is also an expansion of the 'light hypothesis' partition to include the STR versions of the (speed-of-)light hypotheses. As a result, the addition of the STR has no bearing on the original (speed-of-)light hypotheses, in conformity with Reverse Bayesianism. It might be added that, if the new propositions of which the agent becomes aware *were* apparently evidentially relevant to the basic propositions in the old awareness context, then we would *not* have a case of *genuine* awareness growth, to which Reverse Bayesianism is limited.[36]

---

[36] The implication is that we would rather have a case of irrational and/or poorly represented belief change.

This way of saving Reverse Bayesianism however seriously weakens the commonsense appeal and normative interest of the thesis and seems rather *ad hoc*, as the examples under consideration are surely as genuine cases of awareness growth as any. Moreover, if the aim is to represent the internal perspective of an agent, then it is surely more natural to take the individuation of propositions at face value, such that, with respect to our example earlier, the speed-of-light hypothesis corresponds to the same proposition before and after recognition of the Special Theory of Relativity. But that means that new propositions may well have a bearing on the relative probabilities of old basic propositions. Better to modify the Reverse Bayesian principle itself than to modify what counts as genuine awareness growth.

So, we can conclude that we should not impose Reverse Bayesianism as a general constraint on how a rational agent can revise her credences when her awareness grows. The above counterexamples, however, both involve what we called awareness growth by *expansion*. But as previously mentioned, proponents also want to impose Reverse Bayesianism as a constraint on how a rational agent can revise her credences when her awareness grows due to *refinement* (see e.g. Karni and Vierø 2013: 2803). And one might well hope that, despite the mentioned counterexamples, the principle could be retained for belief revision due to refinement.

Unfortunately, counterexamples similar to those discussed also undermine Reverse Bayesianism understood in this latter way. Consider a third example:

Example 3. Suppose you are deciding whether to see a movie at your local cinema. You know that on the day in question, the cinema only shows 'international' (non-English) movies. You realise that both the movie's language and genre will affect your viewing experience. The possible languages you consider are French and German and the genres you consider are thriller and comedy. But then you realise that, due to your poor French and German skills, your enjoyment of the movie will also depend on the level of difficulty of the language. Since you know the owner of the cinema to be simple-minded, you are, after this realisation, much more confident that the language of the movie will be simple than that it will be difficult. Moreover, since you associate simple language with thrillers, this makes you more confident than you were before that the movie on offer is a thriller as opposed to a comedy.

The important feature of this example is that the original awareness context is partitioned according to some property (the language level) that is taken to be evidentially relevant to the comparison of some pair of inconsistent basic propositions – that the movie is a thriller and that it is a comedy – in the old awareness context.

There is one remaining potential objection to, in particular, examples 2 and 3 that we should address before proceeding. The objection is that, in example 2, you realise that you did not, to begin with, have as much evidence as you had thought for your partner having an affair, whereas in example 3 you realise that you had evidence all along that the movie would be a thriller. So, in both cases, you realise that your previous epistemic state was flawed.[37]

Now, we need not reject this characterisation of these examples for them to be counterexamples to Reverse Bayesianism. After all, the reason you realise that your previous epistemic state was flawed is that you become aware of new contingencies that show that you either had more or else less evidence than you thought. In the movie example you become aware of a fact that connects what you already knew to the movie being a thriller, whereas in the surprise party example you become aware of a new possible explanation that convinces you that the evidence you took yourself to have for your partner's affair wasn't strong. So, even though you do realise that your previous epistemic state was flawed, you do so by becoming more aware, and this growth in awareness can be rational without aligning with the requirements of Reverse Bayesianism. In addition, and more generally, one might of course wonder why the kind of conservativity that Reverse Bayesianism captures should be rationally required when one has realised one's previous epistemic state to be flawed.[38]

In sum, the examples show that Reverse Bayesianism cannot hold in full generality, neither as a constraint on belief revision due to expansion nor as a constraint on belief revision due to refinement. Before closing, however, we note a different potential criticism of our analysis. It might be argued that our examples are not illustrative of a simple learning event (a simple growth in awareness); rather, our examples illustrate and should be expressed formally as complex learning experiences, where first there is a growth in awareness, and then there is a further learning event that may be represented, say, as a Jeffrey-style or Adams-style (recall footnote 32) learning event.[39] In this way, one could argue that the awareness-growth aspect of the learning event always satisfies Reverse Bayesianism (the new propositions are in the first instance evidentially irrelevant to the comparison of the old basic propositions). Subsequently, however, there may be a revision of probabilities over some partition of the possibility space, resulting in more dramatic changes to the ratios of

---

[37] Thanks to Teru Thomas for suggesting this interpretation.
[38] Thanks to Alan Hájek for this last point.
[39] This suggestion resonates with the discussion in Hill (2010).

probabilities for the old basic propositions. The reason we reject this way of conceiving of the learning events described by our examples is that the two-part structure is ultimately unmotivated. The second learning stage would be rather ad hoc, since engineered to conform to (a generalisation of) Bayesian conditionalisation. Hence, this would again seem to us to be an artificial and *ad hoc* way to save Reverse Bayesianism.

4.5 Concluding Remarks on Section 4

In this section we set out to determine the norms for belief revision under growing awareness that are the suitable parallel to or extension of the traditional Bayesian norms for belief revision. We have seen that the arguably most-worked-out proposal to this effect, namely Reverse Bayesianism, does not hold in general. That is, there are examples where an agent's awareness grows in a way that conflicts with Reverse Bayesianism, without any apparent irrationality or undue lack of conservatism on behalf of the agent. The further question, to be pursued in the next section, is whether there is some alternative norm of belief change for growing awareness. More generally, what are the features, if any, of conservative belief change when one becomes aware of new contingencies?

5 Awareness Rigidity

5.1 Introduction

In the last section we argued that Reverse Bayesianism fails with respect to propositions A and B if the awareness growth favours one of these propositions over the other. What happens in these cases is that the 'new' propositions that the agent comes to be aware of change the relationships between the 'old' propositions. Figure 2 provides a stylised (*not to scale!*) pictorial illustration of this phenomenon with respect to the Special Theory of Relativity example of awareness growth by expansion. The figure makes clear that the propositions concerning whether or not the speed of light is relative to the observer, denoted 'Relative Light' and 'Non-relative Light' respectively, are associated with different sets of possibilities when awareness about fundamental scientific theories grows to include the Special Theory of Relativity, 'STR'.

Figure 2 moreover suggests a retreat from Reverse Bayesianism to the kind of rigidity principle defenders of Reverse Bayesianism apparently take as fundamental. (Recall the quotes from Section 4.3 which reveal that defenders of Reverse Bayesianism take this principle to be the consequence of appropriate conditional attitudes remaining constant or rigid under growing awareness.) Informally, the idea is that the probabilities of the old propositions, conditional on, roughly speaking, 'how things were before' (in our example, the proposition

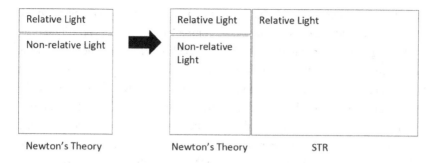

Figure 2 Expansion to include STR

'Newton's Theory'), should be rigid or unaffected by the awareness growth (in our example, the expansion of the fundamental theory space to include 'STR'). Such a rigidity principle looks to be the appropriate basis for conservative belief change in the awareness-growth setting.[40] It is just that the condition does not entail Reverse Bayesianism when stated in general terms, namely, for all pairs of basic propositions. Or so the argument might go.

Even if this position were roughly right, the relevant rigidity condition would need to be properly spelled out. For instance, what precisely is meant by 'how things were before', which should be conditioned on according to this view? In Section 5.2, we appeal to the model we introduced in Section 3.3 in order to meet this challenge. We give a precise definition of the intended rigidity principle, which we dub *Awareness Rigidity*. In Section 5.3, we go on to argue against Awareness Rigidity. The prospects for a general principle for belief change under growing awareness are thus dim. Nonetheless, we proceed, in Section 5.4, to formulate what we call *Restricted Reverse Bayesianism (RRB)*. Even if RRB is not a substantive norm for belief revision under growing awareness, it may yet serve as a guide for detecting cases where we can expect the type of conservatism enshrined in Bayesianism to hold.

5.2 Awareness Rigidity Defined

Let us start then by spelling out the rigidity condition that plausibly constrains belief change under growing awareness. The hope would be that this rigidity condition does not imply Reverse Bayesianism in the situations in which we

---

[40] In fact, Karni and Vierø (2013) formulate the preference analogue of such a rigidity principle, which they call *Awareness consistency*, and use it in their derivation of Reverse Bayesianism. (Karni and Vierø 2015 however use a somewhat weaker axiom.) Moreover, as we will point out, Bradley (2017) explicitly assumes what is essentially Awareness Rigidity (although giving it a different name).

do not want the latter to hold, that is, in situations where the awareness growth is intuitively evidentially relevant to the comparison of some propositions.

Our suggestion for specifying a rigidity condition for awareness growth is to identify the smallest set of possibilities in the new awareness context that is associated with a proposition that used to be associated with all possibilities (i.e., the tautology) in the old awareness context. (Recall the terminology we introduced in Section 3.3.2 to describe the possibilities associated with any given proposition in a particular awareness context.) For the example described in Figure 2, this proposition will indeed be 'Newton's theory', as per our earlier informal discussion. With respect to our rent-or-buy example, when awareness grows by expansion to incorporate the new contingency 'Apartment is sold', as per the shift from Table 11 to Table 12, the proposition associated with the smallest set of new possibilities and which used to be the tautology in the old awareness context is the disjunction of all the old landlord contingencies, i.e., 'Available at higher rent ∨ Available at same or lower rent'. In special cases of awareness growth by refinement where *all* possibilities are effectively refined (what we refer to as *complete* refinement), the relevant proposition will be associated with the entire set of possibilities constituting the new awareness context. So, for instance, when awareness grows by refinement to incorporate the new contingencies concerning the interest rates, as per the shift from Table 12 to Table 13, the proposition associated with the smallest set of new possibilities and which used to be the tautology in the old awareness context is, for instance, 'Available at higher rent ∨ Available at same or lower rent ∨ Apartment is sold'. This proposition is the tautology – associated with all possibilities – in the new awareness context too.

More formally: let T^*_{X+} in $\mathcal{F}_{X+}$ be the proposition(s) found by the following procedure. First, find all those propositions in $\mathcal{F}_{X+}$ that are associated with the full set $\mathbf{W_X}$. (Recall from Section 3.3.2 that the set of possibilities associated with a proposition are just all those possibilities for which that proposition is true.) This results in a set of propositions that we can denote $\{T^*_{X+}\}$. Next, find a proposition in $\{T^*_X\}$ that corresponds to the smallest subset of $\mathbf{W_{X+}}$. Any such proposition is denoted T^*_{X+}. This allows us to specify a rigidity condition that one might take to be the appropriate extension of Bayesian (i.e., conservative) belief change to the case of growing awareness:

Awareness Rigidity. *For any rational agent and for any $A \in \mathcal{F}_X$:*

$$P^+(A \mid T^*_{X+}) = P(A).$$

This, we think, captures the rigidity condition that defenders of Reverse Bayesianism take as more fundamental than Reverse Bayesianism itself. In fact, setting aside subtle differences when it comes to the interpretation of the background algebra (in particular, whether parts of it remain constant during awareness change, which we deny), this is precisely the norm that Bradley (2017: 258) endorses as the appropriate extension of Bayesian conservatism to situations where awareness changes.[41]

Figure 2 makes vivid that Awareness Rigidity does not generally entail Reverse Bayesianism, at least for cases of awareness growth by expansion. Awareness Rigidity requires that the probabilities for old propositions – 'Relative Light' and 'Non-relative Light' – *conditional on 'Newton's theory'*, remain constant when awareness grows. One can see just by looking at the figure that it does not follow that the ratio of the absolute probabilities for 'Relative Light' and 'Non-relative Light' remain constant when awareness grows. For this particular case of awareness growth, then, Awareness Rigidity does seem to precisely capture what aspect of the agent's credences should stay constant as awareness grows.

Note that we might have otherwise defined Awareness Rigidity in a different but logically equivalent way, by concentrating on those propositions that are associated with the agent's atomic possibilities in the old awareness context. Awareness Rigidity requires that the relative probabilities of these propositions remain constant. For instance, with respect to the awareness growth by expansion described in Figure 2, the original possibilities are associated with the propositions 'Newton's Theory & Relative Light' and 'Newton's Theory & Non-relative Light'. It is the ratio of probabilities for these propositions that must remain constant, according to Awareness Rigidity, when the agent becomes aware of a further fundamental theory: 'STR'.

5.3 Against Awareness Rigidity

Unfortunately, Awareness Rigidity is not a compelling rationality requirement, especially for cases of refinement, which we turn to now. The problem is that Awareness Rigidity entails Reverse Bayesianism in those special cases where awareness grows by refinement of *all* possibilities, as per the shift from Table 12 to Table 13 (since it effectively requires that the probabilities for all propositions in the old awareness context remain unchanged). We previously argued that Reverse Bayesianism is not plausible even in these special cases of complete refinement. By *modus tollens* Awareness Rigidity is then not a plausible principle for belief change.

---

[41] Bradley however uses the term 'rigid extension' rather than Awareness Rigidity. More precisely, he calls one (more aware) probability function a rigid extension of another if the two are related by what we call Awareness Rigidity.

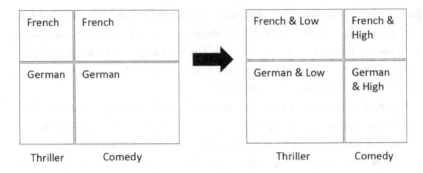

Figure 3 Refinement into high/low language

Is there any reason to retract this position? We think not. Figure 3 is a pictorial representation of our third counterexample to Reverse Bayesianism – the refinement-counterexample to Reverse Bayesianism concerning the international movie that will be shown at the local cinema. Figure 3 makes vivid how awareness of an entirely new property or kind of contingency may intuitively cause adjustment of the relative probabilities of 'old' propositions. Specifically, the relative probability of the thriller and comedy genres changes when the agent becomes aware of the language-level contingencies, denoted 'High' and 'Low'.

While we hesitate to read too much into a formal representation of awareness growth, we suggest that our model introduced in Section 3.3 offers some explanation for why Awareness Rigidity may be reasonably violated. In short, the set of possibilities associated with any proposition completely changes when awareness grows. Initially the set is constituted by possibilities that are effectively truth functions over a given set of basic propositions; after awareness growth, the set is constituted by possibilities that are truth functions over an enlarged set of basic propositions. So, it is not surprising that the relationships between propositions, even conditional on 'how things were before', may in some cases dramatically change. This is what happens in the movie example depicted in Figure 3: when the set of possibilities associated with the movie genres 'Thriller' and 'Comedy' change, so too do their probabilities conditional on 'how things were before', which, in this case, equate to their unconditional probabilities. Examples like this highlight that there is no clear delineation between those propositions and corresponding credences that are and are not affected by the awareness growth.

Mahtani (2020) presents an example that also serves as a striking counterexample to Awareness Rigidity in the case of refinement.[42] The example goes as follows:

> You know that I am holding a fair ten pence UK coin which I am about to toss. You have a credence of 0.5 that it will land HEADS, and a credence of 0.5 that it will land TAILS. You think that the tails side always shows an engraving of a lion. So you also naturally have a credence of 0.5 that [LION] it will land with the lion engraving face-up: relative to your state of awareness, TAILS and LION are equivalent. [...] Now let's suppose that you somehow become aware that occasionally ten pence coins have something other than a lion engraving on the tails side. In particular, you become aware that [STONE] there are some ten pence coins that have an engraving of Stonehenge on the tails side. Let's assume that no other possibilities occur to you.

The awareness growth in Mahtani's example is one of refinement (albeit *partial* refinement as opposed to *complete* refinement, since not all possibilities are refined): you realise that TAILS can be further refined into the disjunction LION ∨ STONE. To see that this is a counterexample to Awareness Rigidity, note that the smallest set of possibilities in the new awareness context that is associated with what used to be the tautology in the old awareness context is LION ∨ HEADS. So, according to Awareness Rigidity, we should find that, say, $P^+(\text{LION} \mid \text{LION} \vee \text{HEADS}) = P(\text{LION}) = 0.5$. But that is counterintuitive. By learning that some instances of TAILS are not instances of LION, and nothing more, your credence in LION should diminish. In other words, intutively, we should find that $P^+(\text{LION} \mid \text{LION} \vee \text{HEADS}) < P(\text{LION})$, in violation of Awareness Rigidity. So, we have another refinement-counterexample to Awareness Rigidity.

Awareness Rigidity might nevertheless be compelling for awareness growth by expansion. After all, when awareness grows by expansion, some old propositions are no longer equivalent to the tautology, and hence, Awareness Rigidity does not entail Reverse Bayesianism in cases of expansion. Perhaps Awareness Rigidity is plausible when restricted to expansion, even if Reverse Bayesianism is not. The example involving an expansion of fundamental theories depicted in Figure 2 suggests as much.

Even in cases of expansion, however, it may be that the gained awareness shakes things up sufficiently in one's old awareness state that Awareness

---

[42] Mahtani proposed the example rather as a counterexample to a restricted version of Reverse Bayesianism that we proposed in earlier draft work. We follow up on this in the Section 5.4 (in particular, see our reference to Mahtani in footnote 45).

Rigidity is violated. Again, our formal model underscores the fact that *all* propositions are in some way affected by a growth in awareness, whether it be awareness growth due to expansion or refinement. So it is certainly not off the table that Awareness Rigidity may be violated in cases of awareness growth by expansion. We leave as an open question whether that is something that can rationally happen.

5.4 *Restricted* Reverse Beyesianism

Finally, we note that while Reverse Bayesianism must be abandoned as a general principle of belief revision for growing awareness, it is arguably still an interesting relation that *may sometimes* hold between pairs of propositions in the transition from one belief state to another. As we have seen with the earlier rent-or-buy examples, there are circumstances where Reverse Bayesianism does intuitively hold. We suggest that these are cases in which what is learnt is *evidentially irrelevant* to the pairs of propositions at issue. In other words, Reverse Bayesianism is a useful relationship for *characterising* cases of awareness growth that is evidentially irrelevant for the pair of propositions at issue.

One might say that a restricted version of Reverse Bayesiaism holds. The relevant principle can be stated as follows:

Restricted Reverse Bayesianism (RBB). *For any $A, B \in \mathbf{X}$ (where $P(A) > 0$ and $P(B) > 0$), if*

- *the awareness growth from $\mathbf{X}$ to $\mathbf{X}^+$ is evidentially irrelevant for A vs. B,*
- *and Awareness Rigidity holds 'locally' with respect to A and B,*[43]

then according to any rational agent:

$$\frac{P(A)}{P(B)} = \frac{P^+(A)}{P^+(B)}.$$

Restricted Reverse Bayesianism is not intended to be a substantive normative principle, since the antecedent conditions – stipulating 'local' Awareness Rigidity and evidential irrelevance – secure the conclusion as a matter of logic (we illustrate this later in this section after we formally define evidential irrelevance). RRB is nonetheless supposed to be illuminating in that it

---

[43] That is, the probability of both A and B conditional on 'how things were before' (as spelled out in the previous section) remains constant under the awareness growth from $\mathbf{X}$ to $\mathbf{X}^+$.

reveals sufficient conditions for Reverse Bayesianism to hold for some pair of propositions under awareness growth.

Of course, whether or not it really is the case that Reverse Bayesianism is entailed by the conditions stated in RRB depends on what it means for awareness growth to be 'evidentially irrelevant for A vs. B'. Intuitively, an awareness growth experience is evidentially irrelevant for A vs. B if the awareness growth neither favours A over B nor vice versa. Note that awareness growth can thus be evidentially irrelevant for A vs. B even if it is evidentially *relevant* for both A and B. For instance, an expansion that lowers the probability of A and B by 10% each, say, favours neither over the other but is (negatively) evidentially relevant for both A and B. It is also worth noting that an awareness growth that increases the probability of $A\&B$ without increasing the probability of either A or B counts as growth that is evidentially irrelevant for A vs. B.[44] Since the purpose of the definition is simply to build a sufficient condition where Reverse Bayesianism holds, and since we think that Reverse Bayesianism should hold (for A and B) in these sorts of cases, we take this feature to be a strength rather than a weakness of our definition.

More formally, we propose the following account of evidential irrelevance, *at least for special cases of awareness growth* (more on this restriction shortly):

Definition (Evidential irrelevance). *For any $A, B \in \mathcal{F}_X$, we say that an agent's awareness growth, from awareness context $\mathbf{X}$ to $\mathbf{X}^+$, where $\mathbf{X_j}$ is the set of all basic propositions $X_j \in \mathbf{X}^+$ such that $X_j \notin \mathbf{X}$, is evidentially irrelevant for A vs. B whenever:*

$$\text{either} \quad \text{(i)} \quad P^+(A \mid \bigvee_{X_i \in \mathbf{X_j}} X_i) = 0 = P^+(B \mid \bigvee_{X_i \in \mathbf{X_j}} X_i)$$

$$\text{or} \quad \text{(ii)} \quad \frac{P^+(A \mid \bigvee_{X_i \in \mathbf{X_j}} X_i)}{P^+(B \mid \bigvee_{X_i \in \mathbf{X_j}} X_i)} = \frac{P(A)}{P(B)}.$$

We said that this definition only works for special cases of awareness growth. The special cases are those of pure expansion or pure (and complete) refinement. In these cases, the 'new' propositions of which the agent is aware (those in $\mathbf{X_j}$) are mutually inconsistent, and we suggest our definition of evidential irrelevance only for such cases. For mixed cases of awareness growth, RRB simply will not be illuminating in the intended way, since evidential irrelevance

---

[44] We thank Sven Ove Hansson for pressing us on this point.

must itself be defined in terms of whether or not Reverse Bayesianism holds for the propositions in question.[45]

Let us now briefly illustrate how, given our definition of evidential irrelevance (and assuming 'pure' awareness growth as previously defined), Reverse Bayesianism logically follows from the two conditions in RRB (i.e., from local Awareness Rigidity plus evidential irrelevance). Note first the new (more aware) space of possibilities, W_{X+}, can be partitioned into two sets: on the one hand, the possibilities associated with 'how things were before', T^*; and on the other hand, the possibilities exclusively associated with (the disjunction of propositions in) the set of newly discovered basic propositions, X_j. (Note that in the case of refinement this second set will simply be empty since the possibilities associated with 'how things were before' amount to the full set of possibilities W_{X+}.)

Suppose now that in the shift from W_X to W_{X+}, Awareness Rigidity is satisfied for both A and B, that is, $P^+(A \mid T^*) = P(A)$ and $P^+(B \mid T^*) = P(B)$; hence,

$$\frac{P^+(A \mid T^*)}{P^+(B \mid T^*)} = \frac{P(A)}{P(B)}.$$

For evidential irrelevance, as previously defined, there are two cases to consider, (i) and (ii). Let's first suppose that in addition to the assumed local Awareness Rigidity, (i) holds in the awareness growth from W_X to W_{X+}. That means it is a case of expansion where neither A nor B is consistent with any of the new possibilities, X_j. So, from local Awareness Rigidity it then follows that:

$$\frac{P^+(A)}{P^+(B)} = \frac{P(A)}{P(B)}.$$

Now suppose instead that in addition to the assumed local Awareness Rigidity, (ii) holds in the awareness growth from W_X to W_{X+}. If it is a case of expansion, that means that the ratio of the probability of A to the probability of B conditional on the possibilities associated with X_j (i.e. all those possibilities not associated with T^*) is the same as the ratio of A and B in the old awareness context. (In the case of refinement, evidential irrelevance due to ii) is entailed by local Awareness Rigidity; moreover, there are no remaining possibilities to consider that are not associated with T^*.) So, again, it then follows that:

---

[45] We are indebted to the work of Anna Mahtani (2020) for inspiring this presentation of RRB as a statement of sufficient conditions for Reverse Bayesianism, rather than a substantive norm. Mahtani's counterexamples to a version of RRB we proposed in previous draft work prompted us to refine the principle. The previous version did not make all the requisite antecedent conditions explicit in the statement of RRB.

$$\frac{P^+(A)}{P^+(B)} = \frac{P(A)}{P(B)}.$$

In sum, the two conditions in the statement of Restricted Reverse Bayesianism – that is, local Awareness Rigidity and evidential irrelevance – logically imply the conclusion in the principle – that is, Reverse Bayesianism. Hence, Restricted Reverse Bayesianism cannot, as we previously noted, be a *substantive normative* principle. However, it is, we think, illuminating in that it states sufficient conditions for Reverse Bayesianism (for 'pure' expansion/refinement cases).

Note that evidential irrelevance does not hold in the counterexamples to Reverse Bayesianism discussed in Section 4.4 (these being cases of pure expansion/refinement as required). For instance, the probability ratio (in terms of probability function P^+) of your partner having an affair with your friend to their being warm friends, conditional on your partner and your friend meeting to organise a surprise party, is lower than the 'old' probability ratio (in terms of probability function P) of your partner having an affair with your friend to their being warm friends. Similarly, in the example involving refinement, the probability ratio (in terms of P^+) of the movie being a thriller to its being a comedy, conditional on the disjunction of low- and high-level language, differs from the 'old' probability ratio (in terms of P) of the movie being a thriller to its being a comedy. In sum, both in the counterexamples to Reverse Bayesianism involving expansion and in the counterexamples involving refinement, evidential irrelevance is violated.

On the other hand, the more standard examples of awareness growth, such as the rent-or-buy examples, *are* plausibly cases of evidential irrelevance (at least with respect to some pairs of 'old' basic propositions), according to our definition. In the expansion case, you become aware that the apartment could be sold. This is irrelevant, for instance, to your relative confidence in the neighbours being noisy vs. quiet, since your new probability ratio (in terms of P^+) of the neighbours being noisy to their being quiet, conditional on the rental apartment being sold, is intuitively the same as your old probability ratio (in terms of P) of the neighbours being noisy to their being quiet. Likewise for the refinement case: the realisation that interest rates could go up or else go down or stay the same is irrelevant, for instance, to your relative confidence in the neighbours being noisy vs. quiet, since your new probability ratio (in terms of P^+) of the neighbours being noisy to their being quiet, both conditional on interest rates either going up or else going down or staying the same, matches your old probability ratio (in terms of P) of the neighbours being noisy to their being quiet.

5.5 Concluding Remarks on Section 5

We take our findings to suggest that there is no general, conservative norm of belief revision for awareness growth. (Alternatively – recall our discussion in Section 4.2.2 – one can translate our conclusion as follows: there is no general account of what it means for credences to be *stable* before and after awareness growth.) Reverse Bayesianism simply characterises those cases in which belief revision under awareness growth happens to be conservative in the familiar way, since the relevant beliefs are not affected by the awareness growth. Restricted Reverse Bayesian describes sufficient conditions for when that is the case.

Now, we have, admittedly, only considered two conservative norms of belief revision for awareness growth: Reverse Bayesianism in Section 4 and Awareness Rigidity in Section 5. To understand why we are skeptical of there being *any* fully general, conservative norm of belief revision for awareness growth, recall that being conservative when revising one's beliefs means that one holds certain aspects of one's credences fixed throughout the revision. We take our discussion so far to have established that when awareness grows, there is no clear boundary between those credences that are directly affected by the learning experience and those that are not. This makes the prospects of finding a general conservative norm of belief revision for awareness growth very bleak.

The spirit of our discussion has been simply to assess whether a rational agent would endorse specific principles for belief revision in response to learning. Diaconis and Zabell (1982: 827) refer to this as a 'subjective' mode of justification. They have this mode of justification in mind when stating (e.g., ibid.: 822) that standard conditionalisation and its Jeffrey variant are applicable *just in case one judges that Rigidity would hold in the given circumstances.* We have effectively claimed that a rational agent may reasonably judge that neither Reverse Bayesianism nor Awareness Rigidity would hold in various cases of awareness growth.

Of course, there are also 'stricter' justifications of conditionalisation, at least when it is interpreted as a *planning* norm,[46] including the so-called 'diachronic Dutch book argument'. On this argument, *planning* to revise one's beliefs contrary to conditionalisation in the appropriate learning circumstances has bad pragmatic consequences that should be avoided; specifically, it makes one vulnerable to sure loss. One might wonder whether, notwithstanding our

---

[46] Again, recall our discussion in Section 4.2.2 of the different interpretations of norms of belief revision.

counterexamples, there is an argument of this sort for Reverse Bayesianism, or perhaps for Awareness Rigidity.

The answer is clearly 'no', we say, if the aim is to justify a very general and widely applicably norm for belief revision under growing awareness. As discussed in Section 4.2.2, in the context of growing awareness, *planning* norms have limited viability. An agent cannot always anticipate that she might experience awareness growth, let alone all the details of this awareness growth. So it would seem that she cannot, at least not always, specify a *plan* for belief revision under growing awareness.

We are yet to fully investigate circumstances in which awareness growth *is* anticipated, however. In these special circumstances, there may be a sense in which an agent can specify a plan for belief revision. One might wonder whether some sort of Dutch book argument could be made in favour of a norm for belief revision when it comes to these special circumstances. That is the question to which we turn in Section 7, after having introduced anticipated awareness growth in the next section.

6 Anticipating Awareness Growth

6.1 Introduction

In previous sections, we introduced the general idea of limited awareness and changes in awareness and explained why these are both common and practically important. We have moreover proposed a general model of limited awareness and awareness growth. Finally, we have argued, against most of the previous literature on limited awareness, that there are no general norms of belief revision for growing awareness; indeed, the notion of conservative belief change is not well defined when it comes to awareness growth.

For the purposes of *decision-making*, however, what is of interest is neither limited awareness nor awareness growth in general but rather the *recognition* of one's limited awareness and how it may change over time. If you are completely 'unaware of your unawareness' in some choice situation – that is, if you do not even realise that you *could be* unaware of something that might turn out to be important for the outcome of your decision – then this lack of awareness will not play any role in your deliberation. In contrast, if you find yourself in a situation of 'conscious unawareness' (Walker and Dietz 2011) – that is, you suspect that you lack awareness and may potentially experience awareness growth – then this suspicion could, and arguably often should, play a role in your deliberation.

Situations where one is aware that one has limited awareness that is relevant to one's choices are not uncommon. These will often be, and are usefully conceived as, situations where one anticipates experiencing awareness growth sometime in the future. (Recall that 'planned awareness growth' was one kind of feedback from the world captured in Table 5.) Of course, there will be situations where one does not anticipate such growth but rather predicts that one's limited awareness will persist over time. Even then, however, we suggest that being consciously unaware means that one is at least open to later experiences of awareness growth.

Intuitively, it would be reasonable to anticipate or at least be open to the possibility of awareness growth in *novel* situations – those in which one's own action is unprecedented or in which one may otherwise experience the world in a new way. Moreover, one's past experiences can serve as a guide to novelty. One may come to recognise situations that are in some important respects similar[47] to situations in which one has previously proven to have limited awareness, presumably brought to light by a later experience of awareness growth.

It would also seem reasonable to be open to the possibility of awareness growth in decision-situations that exceed some specified level of *complexity*. Again, one's past experience can serve as a guide: if you have proven to have limited awareness in the past – due to later experiencing awareness growth – in choice situations that exceed some level of complexity, then it would seem reasonable to at least be open to the possibility of awareness growth whenever you find yourself in a choice situation that exceeds that level of complexity.[48]

Radical technological change would seem to be a prime example of a novel (and complex) situation where we should be open to the possibility of awareness growth. For instance, when scientists in the late 1960s managed to send information between computers the size of small houses, most people could not foresee that this would ultimately result in people walking around with pocket-sized computers with access to the near totality of humanity's knowledge. To take another example: with the benefit of hindsight we see that early industrialists were unaware of the possibility that their employment of new manufacturing and transportation techniques would eventually change the climate. Other similar scenarios lead us to think that deployment of a new technology is prone to limited awareness that may later be 'filled in' and moreover turn out to be relevant to the choice at hand. But there will be more ordinary sorts of situations, too, where one may reasonably anticipate or be open to

---

[47] Although the notion of similarity to past decision situations might seem elusive, and perhaps ill-defined, a formal and sophisticated decision theory based on this notion has been developed: Gilboa and Schmeidler (1995, 2001).

[48] This point is made by Grant and Quiggin (2013b), who consider more precise measures of complexity and develop a game theoretic model with unawareness and (what we call) the anticipation of awareness growth.

awareness growth. For instance, one might anticipate becoming aware of new artists or genres when one visits an arts festival – an example to which we shall later return.

In this section we however foreground the more dramatic or extreme cases of 'conscious unawareness' and associated anticipation or openness to awareness growth, such as the prospect of geoengineering that we introduce in Section 6.2. We go on to propose, in Section 6.3, how to model an agent who is open to the possibility of awareness growth, such that this may play a role in her decision-making. Finally, in Section 6.4, we reflect on how our model of openness to awareness growth reveals this phenomenon to be in some ways very ordinary, more or less akin to ordinary reasoning about less-than-fully-specified contingencies as described in any standard decision model.

6.2 Example: Solar Radiation Management

Solar radiation management (SRM), also known as solar geoengineering, has been discussed as a most likely very efficient and effective way of combating (and potentially even reversing) climate change. One such technique would consist in injecting reflective aerosol particles into the stratosphere. These particles would then reflect a small amount of inbound sunlight back out into space, thus making the planet cooler than it would otherwise be.

Although the idea behind SRM is partly inspired by the eruption of large volcanoes that naturally blast reflective sulphate particles into the stratosphere, nothing like the scale of SRM that would be needed to combat climate change has ever been tried. Thus, as for instance the Solar Radiation Management Governance Initiative frankly admits, the potential 'side effects' of SRM are not well understood and are in fact mostly unknown.[49]

SRM would thus seem to be an example where we do, and should, anticipate awareness growth, if (or when) it is seriously tried. There may be various possible 'side effects' which we cannot yet articulate, but we think one of these side effects is bound to occur if we do try SRM on the scale needed to combat climate change. So none of the possible outcomes of SRM can currently be described in perfect detail. That is to say, we may anticipate that trying out SRM on a global scale leads to awareness growth by *refinement*.

In addition (or alternatively), we may suspect there is some unknown contingency that is inconsistent with any contingency of which we are aware and that could make solar radiation management either much more positive or much more negative than the possible outcomes of which we are currently aware.

---

[49] See www.srmgi.org/what-is-srm/.

That is to say, we may anticipate or at least be open to awareness growth by *expansion*, if we try SRM on a global scale.

Despite SRM being an example where we anticipate awareness growth (should we try it on a global scale), we may nonetheless have some rough estimate of the desirability of SRM. For instance, although we recognise that things could go wrong in ways we have not yet considered, we might think that the expected utility is positive (when SRM is compared to not employing SRM). That is, we may prefer employing SRM to not employing it, despite recognising our limited awareness and the potential for later awareness growth. More generally, the fact that we anticipate or are open to awareness growth that affects the desirability of an action need not entail that we cannot evaluate the action – and compare it to its alternatives – in terms of (at least rough, or imprecise) expected utility.

It might seem puzzling that we can evaluate an option, by estimating its expected utility, even when we take the option to be associated with potential awareness growth. We have more to say about this in the following sections, but let us here offer some preliminary remarks. To begin with, it helps to see the more dramatic cases of awareness growth at one end of a spectrum that also includes, at the other end, the less dramatic, ordinary cases of awareness growth.

Part of the appeal of going to a music or film festival, for instance, is the prospect of finding out about new artists or genres; that is, the possibility of *becoming aware* of artists or genres the existence of which one had previously been unaware. So, we may hope for some abstract outcome or experience, the content of which we cannot quite articulate yet. Still, it seems evident that we can and do compare the option of going to a film festival, say, with the option of staying home, despite there being important aspects of the film festival of which we realise we are currently unaware.[50] The same arguably applies to the evaluation of SRM, despite it being a more extreme case of anticipated awareness growth.

---

[50] Since discovering a new film genre might be 'transformative', for instance in the (epistemic) sense that one cannot know what experiencing films of that genre is like before having had the experience, our claim that one can compare the option of going to a film festival with the option of staying home is in contrast with Paul's (2014) influential work on 'transformative experience'. This is not the place to discuss Paul's argument in detail, but in a nutshell, our response is that even if one cannot fully know what it is like to, say, experience some art genre before having experienced it, that does not rule out the possibility of evaluating the choice-worthiness of options involving that experience by, for instance, reasoning in accordance with a decision-model like that discussed later in this section. (For a similar response to Paul's argument, see Bykvist and Stefánsson 2017).

Nevertheless, it might seem reckless, especially in extreme cases of anticipated awareness growth, to base one's choices entirely on considerations of expected utility. For instance, even if, conditional on the contingencies of which we are currently aware, SRM has positive expected utility, one might wonder whether that suffices to justify a choice to try SRM. After all, since we anticipate awareness growth, we are open to the possibility that SRM will go wrong in ways that we haven't yet considered. In particular, it might seem that agents who are averse to risk and uncertainty should take such possibilities to be reasons against basing decisions purely on considerations of expected utility maximization.[51]

However, it is unclear whether we should take such caution to be inconsistent with expected utility maximisation. In fact, the model we shall soon introduce makes such caution consistent with expected utility maximisation. In particular, the model allows for maximising expected utility *all-things-considered* while turning down an option that has positive expected utility *given the contingencies of which one is aware*.

6.3 Modelling Anticipated Awareness Growth

Previously we argued that when modelling an agent's limited state of awareness, from her own perspective, it is not always apt to use a 'catch-all' to represent that which she is unaware of. In particular, if the agent is 'unaware of her unawareness', and thus does not anticipate any growth in awareness, then in so far as we are trying to model the agent's deliberation, we should resist that modelling choice. The upshot is that the experience of awareness growth cannot always be modelled as the refinement of a catch-all or some other proposition(s).

But that is not to say that it never makes sense to appeal to a catch-all proposition to describe an agent's reasoning. If we are trying to model an agent who is aware of her unawareness, and in particular one who anticipates awareness growth (by expansion), then it does seem natural to use some sort of catch-all to model that which the agent takes herself to be currently unaware of. However, as the reader may recall from Section 3, by 'catch-all' we (unlike some philosophers whose work we discussed) do not here mean strictly speaking *all* possibilities of which the agent is unaware. That is not something that can sensibly feature in an agent's reasoning, or so we argued. Rather, we take the

---

[51] Similarly, one might speculate that uncertainty-averse agents – who are, to put it roughly, particularly worried about the unknown – might place a higher value on increasing their awareness than those who are less risk averse. In fact, Quiggin (2016) proves that within his framework, the expected value (to an agent) of gaining awareness is greater the more risk-averse is the agent.

Table 14 Decision model with no room for awareness growth

	s_1	...	s_n
f	$f(s_1)$	...	$f(s_n)$
g	$g(s_1)$	...	$g(s_n)$

Table 15 Decision model with room for awareness growth

	E_1	...	E_n	??
f	$f(E_1)$	...	$f(E_n)$	$f(??)$
g	$g(E_1)$	...	$g(E_n)$	$g(??)$

catch-all in question to be some abstract proposition standing in for a broad class of contingencies that the agent thinks she may later be in a position to concretize. For clarity, we will refer to the kind of proposition we have in mind as a 'subjective catch-all'.

It may be useful to start with a standard decision-model (in which there is no room for awareness growth), against which we contrast a model of an agent who is aware of their potential lack of awareness and how it may grow. The matrix in Table 14 is such a model. The agent assumes there are n states of the world (from s_1 to s_n) that determine the outcome of the available acts, i.e., the objects of choice, f and g. Finally, $f(s_i)$ is the outcome that obtains when f is chosen (or performed) and s_i is the actual (or true) state of the world.

Since both the states of the world and the outcomes in this model are thought to be maximally specific in all ways that might be relevant to the decision – as is usually assumed – and since the states s_1 to s_n are thought to exhaust the set of possible contingencies – as is also usually assumed – the model makes no room for awareness growth. For instance, if each outcome $f(s_i)$ is thought to contain no uncertainty whatsoever – which in turn means that each state s_i is taken to be a maximally specific description of ways in which the world might be (prior to the choice between f and g) – and if the agent is in addition certain that one of outcomes $f(s_1)$ to $f(s_n)$ will be realised by the choice of f, then an agent who is appropriately modelled with a matrix like the one in Table 14 takes herself to be fully aware (in this decision situation).

Consider now the matrix in Table 15. We focus initially on the openness to awareness growth by *refinement* that is represented here. Suppose that E_1 to E_n

is the finest and most exhaustive 'partition' of the space of contingencies that the agent of interest can come up with. (We shall later see that it is not really a partition, according to the agent.) Each E_i is not taken to be a (fully specified) state of the world but rather an *event* that may not fully determine the outcome of the two available acts. So, for instance, $f(E_i)$ leaves open what precisely are the concrete outcomes that may arise. The agent recognises this, let us assume, but does not know how to further partition the E_i into elements that leave no room for uncertainty. In other words, the agent realises that she is (or, at least, she takes herself to be) unaware of some contingencies that are consistent with each E_i but which, if they materialise, might deliver different outcomes for the two acts.

Now we turn to the openness to awareness growth by *expansion* that is represented in Table 15. We see from the '??' column in the table that the agent does not regard E_1 to E_n to be an *exhaustive* set of mutually exclusive events. In other words, the agent is aware of the fact that there might be contingencies that are inconsistent with all of E_1 to E_n and which would, if they materialise, determine the outcome of the two available acts. These further contingencies, that the agent cannot properly articulate, are represented in the table by '??', which stands for a subjective catch-all. Note that, while we call this openness to awareness growth by 'expansion', formally speaking, the agent here is open to the possibility that she will *refine* her catch-all. There is a sense in which she cannot truly be open to, let alone anticipate, awareness growth by expansion.[52]

One reason it is important to model openness to (and anticipation of) awareness growth in the way just discussed is psychological realism. That has been our focus so far in this discussion. If the agent anticipates (or at least is open to the possibility of) awareness growth, then a decision model in which her 'states' are somewhat crude, and/or in which there is a variable or catch-all representing contingencies she is not fully aware of, is consistent with how she herself sees her epistemic predicament. (The same is not true when an agent is completely unaware, that is, when she is not even aware of her own potential limitations as far as awareness is concerned.) But it is important then that any such catch-all is a subjective one. No psychological realism is gained by adding propositions to the model that go beyond the abstract contingencies that the agent herself has gotten the whiff of.

---

[52] Recall that in Section 3.4 we mentioned various models in the literature that treat all cases of awareness growth as awareness growth that was anticipated. As per our discussion here, these models appeal to either a 'catch-all' or else to otherwise coarse events that the agent expects will be later refined or made more concrete.

Another reason for pursuing our model of openness to awareness growth is that it will often be necessary for describing (and rationalising) an agent's reasoning about what to do. Consider again the solar radiation management example and suppose that a policy-maker's epistemic situation, including where she thinks there is room for awareness growth, when she deliberates about whether to try SRM, is represented by a table like Table 15. We can read the options f and g as doing and refraining from SRM respectively.

First, let us focus on openness to awareness growth by expansion, represented by the subjective catch-all, '??', which in some sense is easier for a modeller to get a handle on. Suppose that the policy-maker reasons that if any of the events E_1 to E_n obtain, then she would prefer implementing SRM rather than abstaining. Still, because the policy-maker anticipates, or is at least open to the possibility of, awareness growth by expansion if SRM is implemented (as represented by the subjective catch-all), she finds SRM to be so risky that she prefers to abstain all things considered. This is to say that the utility and probability of the subjective catch-all, according to the policy-maker, makes the overall expected utility of SRM lower than the alternative, according to her.

To make this claim more precise, suppose that the policy-maker's preference relation *conditional on* events E_1 to E_n[53] satisfies the appropriate coherence constraints, such that this conditional preference relation can be represented as maximising expected utility (recall our discussion in Section 1.2.2 of representation theorems and our later mention in Section 1.3.2 of Karni and Vierø's (2017) representation theorem for anticipated awareness growth). Then we can, in this case, infer from her preferences that the expected utility of implementing SRM, given the catch-all, ??, is sufficiently negative, compared to the expected utility of not implementing SRM given ??, such that SRM is not worth the risk all things considered. In contrast, had the decision-maker preferred to implement SRM, then, given that she prefers SRM over its alternative conditional on events E_1 to E_n, we would instead have inferred that the expected utility of implementing SRM given ?? is *not* sufficiently negative compared to the expected utility of not implementing SRM given ??. More generally, when we use a model that includes a subjective catch-all, the assumptions of standard decision theory for full awareness (supposing the contingencies of which she is aware) allow us to say at least something about the agent's attitudes to that of which she suspects she currently has limited awareness.

---

[53] Piermont (2017) calls such conditional preferences 'contingency plans', and uses them to formally characterize anticipated awareness growth (or 'introspective unawareness', to use his term).

A method like this cannot, however, be used to estimate the policy-maker's attitudes to contingencies that she anticipates or is open to discovering due to what we called a *complete* (as opposed to partial) refinement. Recall that we assumed that the events E_1 to E_n are the most fine-grained contingencies that the policy-maker can come up with; however, she suspects that these contingencies can all be refined according to some property of which she is currently unaware. Hence, there is no possibility of considering the policy-maker's preferences conditional on what she takes to be fully specified contingencies to determine her attitude to that of which she is yet to become aware.

In contrast, if the policy-maker is open to the possibility of a *partial* refinement – which we have so far mostly set aside for reasons of simplicity – then a method very similar to the one we discussed to elicit the attitude to potential expansion could be used for potential refinement. Let us focus on some given event E_i that the agent anticipates will (or may) be later refined. For our SRM example, this event may be 'SRM causes severe smog'. Now it may be that the policy-maker is unsure of all of the maximally specific ways in which this event may be true. But she may comprehend some of the ways it may be true. For instance, she may regard one of the maximally specific states constituting E_i to be 'SRM causes severe smog but otherwise status quo (with respect to natural systems functioning)' (call this E_{i1}). Assume that she thinks there is some other way that E_i might be true that she cannot quite articulate; it may be described as 'SRM causes severe smog and otherwise not the status quo' (call this E_{i2}). This latter instantiation of E_i is like a local catch-all; that is, it acts as a catch-all within the partition of outcomes that comprise E_i. Then the comparison of the policy-maker's preference between f and g conditional on E_i with her preference between f and g conditional on E_{i1} – and similarly the comparison of the policy-maker's preference between f and g conditional on E_i with her preference between f and g conditional on E_{i2} – will reveal how her optimism or pessimism about the specific outcomes associated with E_i, which she currently cannot articulate, affects her comparison of the two acts, f and g. Moreover, if the policy-maker's preferences satisfy the appropriate constraints (recall again our discussion of representation theorems), then we can even find precisely how (un)favourable she expects to be the outcomes associated with the local catch-all event E_{i2}.

6.4 Not Such Extraordinary Reasoning?

The reader might wonder whether our model of openness to awareness growth reveals that it is not in fact a hitherto neglected aspect of reasoning. The thought might be that our model previously described is no different from a standard model of reasoning. So either openness to awareness growth is unremarkable or else it is something that has been accommodated in decision models all along. While we are, to a great extent, sympathetic to this line, we see the similarities between openness to awareness growth and 'ordinary' reasoning – to be discussed in what follows – as helpful for better understanding the former, rather than reason to dismiss it.

First, let us consider the extent of similarity. Given our favoured (subjective) interpretation of catch-alls, they are, in many ways, just like any ordinary proposition that we use to model agents' practical reasoning. In general, the propositions that are thought to feature in an agent's reasoning are quite abstract and lacking in details. In deciding whether to ride my bike or else drive to work, for instance, I may not anticipate or be open to any awareness growth and yet nonetheless appeal to rather abstract propositions, like 'some unusual road incident occurs'. That is, I do not dwell on the specific ways in which an unusual road incident may occur (whether it be, for instance, a collision between cars or a flock of swooping magpie birds harassing cyclists). I can form a preference for driving rather than riding given merely the abstractly described circumstance, and I can also assign some probability to said circumstance occurring. Indeed, upon reflecting on any ordinary case of reasoning, one might surmise that, when it comes to abstraction, the difference between a 'catch-all' and other propositions that ordinarily feature in decision models is one of degree rather than kind.

Similarly, there seems to be much in common between, on the one hand, the coarse-grained events we introduced to model agents who anticipate (and more generally are open to the possibility of) awareness growth by refinement and, on the other hand, the maximally specific states of the world in traditional decision models. After all, the states in traditional decision models are themselves abstract to varying degrees; they are only specific enough to account for everything that matters to the agent in comparing the options available to her. Coming back to my decision about whether to drive or ride to work: I do not consider, for instance, whether or not my neighbour has her breakfast before 9am or whether or not the school children will be in the mood to wave to me. That is because these finer details about how the world might be do not matter to my decision, given my goals and values.

That said, there is a difference worth acknowledging: the *reasons for abstraction* differ when it comes to modelling openness to awareness growth compared to modelling 'ordinary' reasoning, i.e., reasoning that does not involve recognition of potential awareness growth. In the former case, the agent suspects that she omits contingencies in the course of her reasoning that she may later recognise to be relevant to her decision but which, regrettably, she is currently unaware of. In the latter case, while the agent may recognise that she omits contingencies in the course of her reasoning, she does not consider this regrettable since she does not regard these contingencies to be relevant to her decision (and if she later comes to think that these contingencies are relevant, then she can simply re-evaluate her decision in light of them). Nor does she expect to become aware of any other contingencies that turn out to be relevant to her decision.

The question is what hangs on this interpretative difference. Apparently not so much, since it seems that any given decision model, like that in Table 15, could be read either way. While we have described Table 15 as a model of openness to awareness growth, it may otherwise serve as a model of an agent's reasoning who is not open to the possibility of awareness growth. It often makes sense for a model to include a highly abstract proposition that is similar to a subjective 'catch-all', or to include a coarse-grained partition of events, even though the agent anticipates awareness growth neither by refinement nor by expansion. And this is true even when modelling the reasoning, or 'internal perspective', of an agent, which is the perspective on which we have focused.

One can see this by reflecting on our earlier example of deciding whether to drive or ride to work. Other examples are in this regard even more vivid. For instance, when deciding how to invest one's pension, one need not take detailed account of extreme events – such as a meteorite striking the earth and killing all of humanity – that would affect each investment in the same way. One way to model this reasoning would be to include an abstract and unspecified proposition which in this case would be interpreted as the set of contingencies that yield the same outcome for all the pension alternatives. As far as abstractness is concerned, there may not be much of a difference between such a proposition and a subjective catch-all. Continuing with this example, one also need not, when making this investment choice, consider all possible percentage points by which interest rates might change; instead, one would presumably just consider coarse-grained events such as 'low', 'moderate', and 'high' interest rates.

Intuitively, one does not want to say in this case that the decision-maker realises that she is currently *unaware* of relevant contingencies that she may later come to be aware of. Quite the contrary. It is her considered judgement that the further details of the contingencies in question are not worth dwelling on since they will not affect her choice. However, if she were to come to believe that, say, a meteorite strike could affect the various investments differently, then she could (and should) include this in her reasoning, unlike someone who is truly unaware of this possibility.[54]

We can also use Table 15 to model an agent who suspects that *something* that she deliberately excluded from her deliberation should in fact be included, while not being able to specify further what she wrongly excluded. In that case, the agent's predicament will be much like someone's who anticipates or is at least open to awareness growth. In fact, there is an intuitive sense in which this agent does anticipate awareness growth: she anticipates becoming aware of the fact that something that she took to be irrelevant to her deliberation in fact is relevant.[55]

Finally, we said earlier that the similarities between openness to awareness growth and ordinary reasoning may be instructive. For starters, the similarities provide further grounds for the proposal of Section 6.3 that agents who are open to awareness growth can maximise expected utility just like ordinary reasoners. Furthermore, we have grounds to think that the credences and utilities underlying these expected utility evaluations are largely based on the same sorts of considerations, whether the case involves anticipated awareness growth or not. With respect to the SRM example, for instance, that the policy-maker anticipates later becoming more aware of the contingencies represented by the catch-all does not mean that she cannot evaluate the probability and utility of the catch-all based on what she is currently aware of, sketchy as that may be. For instance, the policy-maker may judge the utility of the catch-all to be negative, based on the fact that in the past the unforeseen outcomes of major technological change have been bad sorts of disruptions to existing natural and social systems. This was the sort of reasoning we described in Section 6.1.

By way of contrast, note that some take the evaluation of utilities and probabilities in cases of anticipated awareness growth to be somewhat problematic or else highly idiosyncratic. Laurie Paul (2014), for instance, is sceptical about an agent's ability to evaluate contingencies that she herself takes herself to be unacquainted with (recall footnote 50) and says that, to the extent the agent can do so, all she can go on is the value of some general experience of discovery (the 'revelatory value', as she calls it). While not similarly sceptical about the evaluation of yet-to-be-articulated contingencies, Karni and Vierø (2017) seem to agree on the last point, suggesting that the agent's evaluation

---

[54] Similarly, if the agent were to come to think that, say, the difference between a 0.02 and 0.021 interest rate is of importance to her decision, then she could factor this into her reasoning, unlike someone who is truly unaware of this possible refinement.

[55] Karni and Vierø (2017: 317) make a similar observation about their model of anticipated awareness growth.

of a catch-all is to be interpreted as how much she generally likes being surprised.[56]

We resist such exceptional treatment of anticipated awareness growth. To be sure, when an agent anticipates awareness growth, this has some bearing on the bases for her belief and desire judgements. For instance, her general attitude towards surprise plausibly plays a greater role than usual (as we suggested with our earlier arts festival example). Moreover, the agent will have different sorts of projections about her attitudes in the future. Precisely how these projections of her more aware future self can and should bear on her present attitudes is the topic of Section 7. For now we emphasise that the reasoning of an agent who anticipates awareness growth – and, more generally, the reasoning of an agent who is open to the possibility of awareness growth – has much in common with what we have called 'ordinary' reasoning.

6.5 Concluding Remarks on Section 6

We have now explained in informal terms what it means to be open to the possibility of awareness growth, and even to anticipate awareness growth, and illustrated why such openness is both common and important. We have more-over suggested a formal way of accounting for this within a decision model. We appealed to abstract propositions, whether a 'subjective catch-all' in the case of expansion, or else a coarse-grained 'partition' of events in the case of refinement. Moreover, we considered why openness to, and even the anticipa-tion of, awareness growth, thus understood, is in some ways unremarkable. We have however not yet discussed in detail how such anticipation – in particular, an agent's projections of her more aware future self – should (or should not) affect her current credences and preferences. That is the topic to which we turn next.

7 Awareness Reflection

7.1 Introduction

In Sections 4 and 5 we considered the dynamics of awareness growth. In par-ticular, we considered what changes in credence are rational when an agent with limited awareness experiences awareness growth (or more minimally,

---

[56] Moreover, Grant and Quiggin (2013a, 2013b) argue that, when it comes to openness to aware-ness growth, probability estimates are necessarily based on *induction* and options are evaluated by *heuristics*, which differ from the standard (deductive and calculative) methods assumed for fully aware agents.

what changes in credence are consistent with the agent having 'stable' credences). We argued that more or less anything goes. There is no general requirement of rationality dictating how you should change your beliefs upon becoming aware of new contingencies. (Likewise, there is no straightforward way to detect whether your credences are stable upon becoming aware of new contingencies.)

In Section 6 we however turned our attention to a special case of limited awareness; namely, when one suspects that there is something of which one is unaware and perhaps even anticipates growth in awareness. As we saw, this special case of limited awareness is both common and tremendously important from a practical point of view.

In this section we focus in particular on what is entailed when an agent anticipates, or predicts (rightly or wrongly), that her awareness will later grow. While not itself a dynamic phenomenon, anticipated awareness growth is intimately related to the dynamics of belief. We ask whether there is a norm of rationality that constrains an agent's credences in these special circumstances. To be clear, this would not be a *diachronic* norm, governing *change* in credence, but rather a *synchronic* norm, governing an agent's credences *at a given time*. We consider both an informal and a formal argument for such a norm, which seems to be very constraining indeed: when you anticipate awareness growth, your current credences should match your expected[57] future credences in the event that you experience awareness growth. Moreover, there is arguably an equally constraining norm on preference: when you anticipate awareness growth, your current preferences should match your predicted[58] future preferences in the event that you experience awareness growth.[59]

7.2 Informal Argument

We will focus on a very simple decision problem to try to get clear on what (if anything) rationality requires of an agent who anticipates awareness growth. Suppose, for instance, that you are trying to decide whether you should go to the beach rather than stay at home and finish the latest Netflix series that you have been binging on. The pleasantness of going to the beach is highly sensitive

---

[57] As will become apparent, we are using 'expected' here in the technical (mathematical) sense. More precisely, the norm requires that the mathematical expectation of your more aware credences, calculated relative to your own prediction about your future credences, should be the same as your less aware credences.

[58] See footnote 68 for an explanation of the connection between an agent's predicted preference and her current expectation of expected utility.

[59] We thank Michael Nielsen for encouraging us to consider 'reflection principles' in relation to growing awareness and thereby inspiring our investigations in this section.

Table 16 Beach or Home

	sunny	**clouded**	**??**
Beach	Beach & sunny	Beach & clouded	Beach & ??
Home	Home & sunny	Home & clouded	Home & ??

to what the weather will be like, of which you are uncertain. (Let us imagine that you do not have access to a trustworthy weather forecast.) Staying at home is the risk-free option, since you take yourself to know, for instance, that you will have a relaxing day on the couch no matter what the weather will be like.

Now, suppose that the only weather conditions that you consider are *sunny* and *clouded*. You suspect, however, that there is some weather condition that you have left out, which may affect the outcome of your decision to go to the beach. (We, the modellers, may see that this weather condition is *misty*.) Your decision problem is represented by Table 16.[60]

To make the example a little more concrete, suppose that your credence that it will be sunny is 0.4, which is also your credence that it will be clouded. Your credence in the (subjective) catch-all weather contingency, denoted '??' in Table 16, is 0.2. Now we can ask: is this probability, 0.2, indicative of (or constrained by) your predicted future credences? Conversely: once you manage to fill in the catch-all, how do you expect that this will change your credences in all three weather contingencies? In other words, now when you are unaware of the content of the catch-all, what probability do you expect that you will assign it when you become aware of its content, and what probabilities do you expect you will then assign sunny and clouded?

One natural answer would be that you should not expect the above growth in awareness to yield any change, positive or negative, in your credences. You may of course believe that this awareness growth *could* affect your credences in some direction. For instance, you may think that you *could* become more confident of the catch-all, but you should then also believe that this awareness growth *could* affect your credences in the 'other direction', such that you are less confident in the catch-all. More precisely, the claim is that there should be no *expected* change of credence after awareness growth, where the expectation is based entirely on your own prediction about your future epistemic state.

---

[60] The example is inspired by Bradley (2017: 254).

What considerations support the claim just stated? Informally, one could argue for the claim by noting that if you *did* expect your credence in, say, the catch-all to change one way or another once you become aware of its content, then you should now change your degrees of belief in that same direction – assuming (as seems reasonable in this case) that you do not predict that you would be any less rational, or less informed, if your awareness were to grow. For instance, suppose that you expect that if you were to become aware of what the catch-all consists of, you would become more confident than you currently are that it will neither be sunny nor clouded. Then it would seem that you should revise upward your current degree of belief that it will neither be sunny nor clouded. You should, as it were, defer to your more aware self. For why would you not trust someone who is exactly like you, epistemically speaking, except more aware than you are?

Now, it could of course be that when it comes to some cases of anticipated awareness growth, one does *not*, in the relevant sense, treat one's more aware self as someone exactly like one's current self except more aware; and hence, one is *not* willing to defer to one's more aware self. How might such cases be distinguished? It cannot be on the basis of the agent's predicted credence change, since we argued in Sections 4 and 5 that there is no general way to discern what is a stable credence change upon growth in awareness. Perhaps then it is on the basis of the agent's deliberative process and whether it is free from what one might call 'epistemic interferences'. Mind-altering drugs would presumably count as an epistemic interference that makes one's future self, more aware as it may be, unworthy of one's current deference. But so too might other strongly disruptive experiences. Imagine for instance that one predicts the experience of going to an arts festival to be so 'epistemically transformative' (Paul 2014) that the belief system that one will have after the experience differs from one's current system in some fundamental ways (that are neither for the better nor the worse). In that case, rationality may not require that one defer to one's future self and thereby match one's credences to one's expected future credences.

Although we do not want to rule out the possibility of awareness growth being transformative or otherwise accompanied by an epistemic interference, we shall set such experiences aside for now and instead focus on what arguably are more typical examples of anticipated awareness growth. In fact, the principles we will discuss explicitly stipulate the awareness growth to which they apply to be free of epistemic interferences, since they only apply to events where the *only* thing that happens is that you gain more awareness. Hence, these principles do not apply to events where you gain more awareness *and* are subject to an epistemic interference.

We can call the principle that the considerations about the beach-or-home example seem to support *Awareness Reflection*, after the traditional Reflection principle (van Fraassen 1984).[61]

Awareness Reflection (Informal version). *For any awareness context and any proposition A (in that context), if you predict that between now and time t the only thing that happens is that you gain more awareness, then your current credence in A should equal your currently expected credence in A for time t (if the latter is well-defined). Conversely, your currently expected credence in A for time t should (if well-defined) equal your current credence in A.*

The qualification that the relevant values be well-defined is due to the fact that the informal arguments only seem to establish that there is something strange with a person who currently has a credence for A that differs from her currently well-defined expectation of credence at time t. We get back to this qualification in Section 7.3.2, when discussing a formal argument for Awareness Reflection.

In addition to it seeming odd, from a purely epistemic perspective, to violate Awareness Reflection, one might suspect that violating the principle can be undesirable from a purely practical point of view. After all, if you violate Awareness Reflection, then you expect that what you are now willing to pay for some bet is more than what you will find the bet to be worth if you gain more awareness. For instance, if you expect that, once you become aware of the content of the catch-all weather contingency, you will be less confident in it being sunny than you currently are, then what you are now willing to bet on it being sunny is more than what you expect you will be willing to bet when your awareness grows. One might suspect that a clever bookie could exploit this discrepancy. That is indeed the case: a so-called *Dutch book strategy* can be employed against you if you violate Awareness Reflection, as we shall soon see.

7.3 Formal Argument

We now consider a formal argument that supports the informal considerations and conclusions of Section 7.2. We start by formalising the norm, Awareness Reflection, that we introduced earlier, and we then propose a so-called 'Dutch book argument' for this norm.

---

[61] Note that even if Awareness Reflection is true, it need not, in any sense, be a fundamental norm of epistemic rationality. Rather, it may simply be an instance of a more general norm of deference to those who are experts relative to your current self. In fact, Brian Hedden (2015) argues that this is true of epistemic reflection principles in general.

7.3.1 Awareness Reflection Formalised

As usual, we will use P to denote your less aware credences. What matters in the case of anticipated awareness growth is what you predict your credences to be in the future, having experienced awareness growth. Let $\mathbf{P}^+$ denote the *proposition* that your credences in your more aware state can be represented by P^+.[62] Assume for now that you do indeed make such a prediction – that you entertain the proposition $\mathbf{P}^+$ in your current awareness context. In fact, in line with our earlier comments, it is important for what follows that the proposition in question, $\mathbf{P}^+$, is slightly more complex: it denotes that your credences in your more aware state can be represented by P^+ *and* you are otherwise just as informed (and just as rational) as you are now when your credences are represented by P. More generally, your deliberative process is free of what we called 'epistemic interferences'.

As a first attempt, we might try to formally present the principle we have been discussing and will now investigate further as:

Awareness Reflection (Formal version). *For any awareness context X and any proposition $A \in X$, and for any rational P and P^+:*

$$P(A \mid \mathbf{P}^+) = P^+(A).$$

Informally, the principle says that the degree to which you should now believe A, given that you will in your more aware state believe A to some particular degree, say r, is that degree, r. (In Section 7.3.2 we consider a weakening of this principle.)

Note that this (conditional) formulation of Awareness Reflection is, under special conditions, logically equivalent to an expectational version that might on the face of it seem to better capture our informal discussion of responding to your expectation about your more aware degrees of belief. Another virtue of the expectation version is that, unlike Awareness Reflection, it does not give the impression of lacking actual-credence guidance. Faced with Awareness Reflection, one might wonder how one can use that norm to guide one's credences, given that one will never become certain that one will have any particular credence function in a more aware state; hence, one will never be in a position to conditionalise on a proposition like $\mathbf{P}^+$.

---

[62] The supposed domain of this probability function, P^+, is simply your current algebra of propositions, which we earlier denoted $\mathcal{F}$. Your anticipated awareness growth (from $\mathcal{F}$ to $\mathcal{F}^+$) amounts, formally, to a *refinement* of your current set of possibilities including the subjective catch-all. As such, your credences over the propositions in the 'coarse' algebra, $\mathcal{F}$, even once you become more aware, should satisfy the probability calculus.

The special conditions mentioned that are crucial for the logical equivalence claim can be spelled out as follows: you are *certain* you will, epistemically speaking, be the same agent after awareness growth, in that you will not be subject to epistemic interferences,[63] and are uncertain only about which of a number of candidate credence functions will represent your more-aware credences. As such, there is some set of credence functions, call them P_1^+ to P_n^+, that are the candidates for representing your more-aware credences. As before, $\mathbf{P}_i^+$ denotes the *proposition* that (you are not subject to epistemic interferences and) your more-aware credences can be represented by P_i^+. Since $\{\mathbf{P}_1^+, \ldots, \mathbf{P}_n^+\}$ is a partition of the set of all possibilities[64] (and assuming that each has positive probability), the *Law of total probability* entails that:

$$P(A) = \sum_{i=1}^{n} P(\mathbf{P}_i^+) \cdot P(A \mid \mathbf{P}_i^+).$$

But then Awareness Reflection implies that:

$$P(A) = \sum_{i=1}^{n} P(\mathbf{P}_i^+) \cdot P_i^+(A). \tag{7.1}$$

Less formally, Awareness Reflection implies that your current credence in A should equal your expected credence in A in your more aware state. Moreover, the expectational formula implies Awareness Reflection. Updating 7.1 on $\mathbf{P}_j^+$ gives us:

$$P(A \mid \mathbf{P}_j^+) = \sum_{i=1}^{n} P(\mathbf{P}_i^+ \mid \mathbf{P}_j^+) \cdot P_i^+(A \mid \mathbf{P}_j^+). \tag{7.2}$$

But if we then assume that one always is certain of one's own credence, that is, $P_j^+(\mathbf{P}_j^+) = 1$, which implies that $P_j^+(A \mid \mathbf{P}_j^+) = P_j^+(A)$, we get:

$$\sum_{i=1}^{n} P(\mathbf{P}_i^+ \mid \mathbf{P}_j^+) \cdot P_i^+(A \mid \mathbf{P}_j^+) = P_j^+(A). \tag{7.3}$$

Combining 7.2 and 7.3, we get Awareness Reflection:

$$P(A \mid \mathbf{P}_j^+) = P_j^+(A).$$

---

[63] This is in line with Briggs (2009), who argues for a qualification of the traditional Reflection principle which can be put as follows: for reflection principles to be generally plausible, we need to assume that the agent, when their credence is represented by P, is *certain* that they are the same agent as the one that will be represented by P^+. (Cf. one of the interpretations of Conditionalization that we suggested in Section 4.2.2 – that this pattern of belief revision is constitutive of being a stable agent when revising one's credences.)

[64] If, contrary to our stipulated assumption, you think it possible that you will *not* be the same epistemic agent after the growth in awareness due to an epistemic interference, then $\{\mathbf{P}_1^+, \ldots, \mathbf{P}_n^+\}$ is *not* a partition of the possibility space, as these propositions are not exhaustive.

So, the conditional version of Awareness Reflection is logically equivalent (under the conditions discussed) to an expectational version, where the latter is more pertinent to one's actual credences at a time, and was thus the focus of our informal considerations in Section 7.2.

7.3.2 A Dutch Book Argument for Awareness Reflection

Return now to the simple (conditional) Awareness Reflection principle. As those familiar with the Dutch book argument for the traditional Reflection principle (due to van Fraassen 1984) might immediately recognise, a Dutch book strategy can be employed against you if you violate Awareness Reflection. A Dutch book strategy is a betting strategy that consists of bets that you consider individually fair, or acceptable, but which nevertheless together ensure that you lose. In other words, no bet in the strategy is unfavourable, as judged by your own degrees of belief, but together the bets ensure your loss. An important premise in so-called Dutch book *arguments* is that being vulnerable to a sure loss is a sign of irrationality. Hence, if Dutch book arguments are generally valid, Awareness Reflection may be a requirement of rationality.[65]

Here we describe a Dutch book argument for Awareness Reflection.[66] The bets have prices and prizes between 0 and 1. These numbers can, for instance, be interpreted as dollars, or units of well-being, or anything else that we can assume to be valued linearly (at least in the zero to one interval).

Suppose that you violate Awareness Reflection by being more confident of A than you expect you will be when your awareness grows; in particular, your current conditional degree of belief in A, given that you believe A to degree r in your more aware state, is some degree greater than r. More formally, $P(A \mid \mathbf{P}_j^+) > r$ even though $P_j^+(A) = r$; in violation of Awareness Reflection. First the bookie offers you the following three bets, each of which you accept, assuming that you use your degrees of belief or credences to evaluate bets by their expected value.

- Bet 1 costs you $P(A \& \mathbf{P}_j^+)$ and pays you 1 if $A \& \mathbf{P}_j^+$ is true but pays 0 otherwise.
- Bet 2 costs you $P(A \mid \mathbf{P}_j^+) \cdot P(\neg \mathbf{P}_j^+)$ and pays you $P(A \mid \mathbf{P}_j^+)$ if $\neg \mathbf{P}_j^+$ but pays 0 otherwise.
- Bet 3 costs you $(P(A \mid \mathbf{P}_j^+) - r) \cdot P(\mathbf{P}_j^+)$ and pays you $P(A \mid \mathbf{P}_j^+) - r$ if $\mathbf{P}_j^+$ but pays 0 otherwise.

---

[65] For a more detailed discussion of Dutch book arguments, see Richard Pettigrew's Element in this series (Pettigrew 2020).

[66] Our formulation is similar to Vineberg's (2011).

Now, if $\neg \mathbf{P}_j^+$, then you win $P(A \mid \mathbf{P}_j^+)$ from Bet 2, which is the sum of what you paid for Bet 1 (which you have lost) and for Bet 2, and you have also lost Bet 3, for which you paid $(P(A \mid \mathbf{P}_j^+) - r) \cdot P(\mathbf{P}_j^+) > 0$; so, you are at a net loss.

However, if $\mathbf{P}_j^+$, then you have won Bet 3, thus gained $P(A \mid \mathbf{P}_j^+) - r$, for which you paid $(P(A \mid \mathbf{P}_j^+) - r) \cdot P(\mathbf{P}_j^+)$; but you have lost Bet 2, for which you paid $P(A \mid \mathbf{P}_j^+) \cdot P(\neg \mathbf{P}_j^+)$. So, from bets 2 and 3 you are at a loss: the combined net outcome from these bets is $rP(\mathbf{P}_j^+) - r < 0$. But Bet 1 is not settled until the truth of A is known. What the bookie now does is to buy from you a Bet 4 that pays him 1 if A but 0 otherwise, exploiting your new degrees of belief; that is, he offers a price of r for Bet 4, which you accept (again assuming that you use your degrees of belief to evaluate bets by their expected value). Then, if A is true, you win Bet 1 but lose this final bet, and the reverse is true if A is false; so, in either case, you end up with a total of $P(\mathbf{P}_j^+)(r - P(A \mid \mathbf{P}_j^+))$ from the four bets. Thus, by the assumption that $P(A \mid \mathbf{P}_j^+) > r$, you are again at a net loss.

An analogous strategy could be used to exploit you if you had instead violated Awareness Reflection by $P(A \mid \mathbf{P}_j^+) < r$ even though $P_j^+(A) = r$.

In other words, whatever happens, you are sure to lose, and the bookie is sure to win, if you use your awareness-reflection-violating degrees of belief to decide which bets to accept. So, in so far as being vulnerable to sure loss, due to your degrees of belief, is a sign that your degrees of belief are irrational, we can conclude that it is irrational to violate Awareness Reflection.

We should note that this Dutch book argument depends on the assumption that $P(A \mid \mathbf{P}_j^+)$ is defined; that is, that it takes *some* value. (Others have noted the corresponding assumption in the Dutch book argument for the traditional Reflection principle, e.g. Briggs 2009.) It will however be undefined if $\mathbf{P}_j^+$ has zero probability, according to the agent, and similarly if at least one of $\mathbf{P}_j^+$ and $A\&\mathbf{P}_j^+$ has *no* probability, according to the agent. Throughout the Element we have been assuming, for reasons of simplicity, that an agent has precise probabilities in those propositions of which she is aware; hence, if $\mathbf{P}_j^+$ or $A\&\mathbf{P}_j^+$ has no probability, according to the agent, then that means that she is not aware of the proposition(s) in question.

The assumption that $P(A \mid \mathbf{P}_j^+)$ is well-defined is however far from being self-evident. Hence, we can either take the assumption as being part of the Dutch book argument, or we can weaken Awareness Reflection to:

Awareness Reflection (Formal, weaker version). *For any awareness context and any proposition A, and for any rational P and P$^+$:*

$$P(A \mid \mathbf{P}^+) \not> P^+(A).$$

So we see that an agent is only subject to the threat of a Dutch book and thus the norm of Awareness Reflection (now assumed to refer to the stronger version) in those circumstances where she both entertains the various candidates for her future more-aware credences and is also certain she will be free of epistemic interferences upon becoming more aware. But would these conditions ever plausibly hold? For one thing, in the context of awareness change, anticipated as it may be, the agent cannot even articulate the events that she may or may not come to learn, which differs from the traditional cases to which Dutch book arguments have been applied. Perhaps this fact could be used to argue that even if Dutch book arguments are valid when it comes to the ordinary Reflection principle and Conditionalisation, the argument is not ever applicable when it comes to Awareness Reflection. By way of response, however, one might point out that the agent need not be able to predict *what* she will become aware of in order to predict *how* this awareness growth may affect her credences in propositions of which she is already aware. The Dutch book argument, and ultimately the norm of Awareness Reflection, is concerned just with the latter kind of prediction.

7.4 Awareness Reflection vs. Reverse Bayesianism

Although Awareness Reflection, when it applies, is in many ways a very strong and very conservative requirement, it turns out that an agent can satisfy the principle without satisfying, or predicting that she will satisfy, the conservative rule we examined in Section 4: Reverse Bayesianism. In this section we illustrate this fact by returning to the decision between staying at home and going to the beach, represented in Table 16. We go on to discuss why this is further reason to doubt that Reverse Bayesianism is a general norm for belief revision under growing awareness.

Recall that we assumed that, in your less aware state, you believe to degree 0.4 that it will be clouded, and you also believe to degree 0.4 that it will be sunny. Now suppose that, in line with Awareness Reflection, you expect that, after your awareness grows, you will believe each of these possibilities to degree 0.4. However, you believe this to be the case because you now (in your less aware state) believe that you will (in your more aware state) *either* believe to degree 0.2 that it will be sunny and to degree 0.6 that it will be clouded *or* believe to degree 0.6 that it will be sunny and to degree 0.2 that it will be clouded, and you now (in your less aware state) find each of these possible future epistemic states to be equally likely and together exhaustive. More formally:

$$P(sunny) = P_1^+(sunny) \cdot P(\mathbf{P}_1^+) + P_2^+(sunny) \cdot P(\mathbf{P}_2^+)$$
$$= (0.6)0.5 + (0.2)0.5 = 0.4.$$

$$P(clouded) = P_1^+(clouded) \cdot P(\mathbf{P}_1^+) + P_2^+(clouded) \cdot P(\mathbf{P}_2^+)$$
$$= (0.2)0.5 + (0.6)0.5 = 0.4.$$

Note that in this case, while you satisfy Awareness Reflection, you believe that when awareness grows, you will either be three times more confident that it will be sunny than clouded, or three times more confident that it will be clouded than sunny. And since we are assuming that these are the only two future epistemic states that you consider possible, you are in fact *certain* that you will be three times more confident in one of these weather contingencies than the other. Before awareness grows, however, you are equally confident in these two weather contingencies. So, you are certain that you will violate Reverse Bayesianism, understood as a diachronic norm. In fact, you *do* violate Reverse Bayesianism, if it is rather understood as a 'planning' norm (recall our discussion in Section 4.2.2), since you predict or plan that your relative credence in clouded versus sunny weather will change one way or the other upon awareness growth.

Note that Awareness Reflection is strictly logically weaker than (the planning version of) Reverse Bayesianism. If one plans or predicts that one's credence change upon awareness growth will conform with Reverse Bayesianism, then one will also satisfy Awareness Reflection. That is because awareness growth that is anticipated is, formally speaking at least, awareness growth by refinement (we mention this in footnote 62 and also in Section 6.3). In cases of refinement, Reverse Bayesianism requires simply that all credences in propositions of which you were already aware (including here the subjective catch-all) stay constant. So if you satisfy (the planning version of) Reverse Bayesianism, you trivially satisfy Awareness Reflection.

While it is a logically weaker norm, in many cases the person who satisfies Awareness Reflection while predicting a credence change in violation of Reverse Bayesianism behaves like a person who predicts that they will not change their beliefs at all when awareness grows, in accordance with Reverse Bayesianism. For instance, suppose again that you satisfy Awareness Reflection by splitting your confidence between two different future epistemic states, as in the previous example. Moreover, suppose that you want to base your choice, partly at least, on what you expect yourself to believe when you gain more awareness. Then if you are what we might call *uncertainty neutral*[67] *with respect to your future beliefs* – in the sense that, in so far as you take your predictions about your future beliefs into account, you only consider your expected future degrees of belief – then you will act just like someone who predicts that

---

[67] Some may be more familiar with the term 'ambiguity neutral'.

their beliefs will not change at all when awareness grows, in accordance with Reverse Bayesianism.

However, a person who is sensitive to their own uncertainty about their future beliefs, will, even though they satisfy Awareness Reflection, often act quite differently from someone whose predictions accord with Reverse Bayesianism. For instance, if you are averse to uncertainty of this kind, then you might not be willing to risk finding yourself in a situation where awareness grows just as you are arriving at the beach in a way that results in you becoming three times more confident that it will be clouded than that it will be sunny; and hence, you will now essentially act as though you were more confident that it will be clouded than that it will be sunny, even though you are actually now equally confident that it will be clouded as that it will be sunny. In contrast, someone whose predictions about their future beliefs accord with Reverse Bayesianism will not behave that way, since they are not uncertain about their future beliefs in clouded versus sunny.

So, Awareness Reflection is strictly logically weaker than (the planning version of) Reverse Bayesianism, and the two norms can have different behavioural implications. What should we conclude from this? We suggest that this observation casts further doubt on Reverse Bayesianism being a general norm of rational belief change for growing awareness (however such a norm is interpreted). Awareness Reflection is highly constraining with respect to how one's predicted changes in credence upon awareness growth should relate to one's current credences. And yet this norm does not require that one's predicted changes in credence accord with Reverse Bayesianism. So the latter norm apparently goes out on a limb. Moreover, the traditional arguments do not seem to offer any support for this limb. For instance, the Dutch book argument proposed in Section 7.3.2 only secures (as far as its assumptions hold) the weaker norm, Awareness Reflection.

7.5 Preference Awareness Reflection

The intuition behind – and the informal argument in favour of – Awareness Reflection can be applied more generally to your preferences: if you predict[68] that as your awareness grows (but everything else remains fixed) you

---

[68] We can think of this prediction of yours as corresponding to your expectation of expected utility. For instance, that you now predict that you will later prefer not to go to the beach means that your current credences over your future credence and utility functions is such that your current expectation for your future expected utility assignment to the option of going to the beach is lower than your current expectation for your future expected utility assignment to the option of not going to the beach.

will reverse a current preference, then that is arguably a reason to reverse it now. For in this case too one could ask why you would not defer to someone who is exactly like you in every respect except that they are more aware than you are.

Consider again for instance the beach-or-home example. Suppose that you predict now that when you are able to specify the catch-all weather contingency, you will prefer not to go (or not to have gone) to the beach. (And as before, let us suppose that you predict that you will be no less rational, and no less informed, when your awareness has grown.) Then it would seem that, intuitively, you should not prefer to go to the beach now, in your less aware state. In other words, it would seem we should accept:

Preference Awareness Reflection. *For any awareness context and any pair of actions a and b, if you predict that, between now and time t, the only thing that happens is that you gain more awareness, then you should not now predict that your preference ranking of a vs. b at time t differs from your current ranking of a vs. b.*

It might be illuminating to compare this principle to a stronger and more general version. Let's use the term 'Preference Reflection' for the principle that you should not now predict that your preference ranking of a vs. b at some later time t differs from your current ranking of a vs. b.[69] Note that Preference Awareness Reflection is a special case of Preference Reflection in that the former says that the latter holds in the special case where the only thing that happens until time t is that the agent gains more awareness. We do not claim that Preference Reflection is a general requirement of rationality. However, it would seem irrational to violate the general principle just because you predict that you will gain more awareness.[70]

Since Preference Awareness Reflection only holds in cases where you predict that 'between now and time t, the only thing that happens is that you gain more awareness', the most obvious complaints one would have about a more general principle like Preference Reflection seem not to apply to this

---

[69] Preference Reflection is similar to what Arntzenius (2008) calls 'Desire Reflection', which is stated in terms of a numerical representation of desire (i.e., utility or 'desirability') rather than in terms of binary preference.

[70] Similarly, Arntzenius claims that Desire Reflection should not be violated merely because one conditionalises upon new evidence (2008: 279). He weakens Desire Reflection by adding a condition stipulating that the only thing that happens in the relevant time interval is that the agent conditionalises on new evidence, and he calls the resulting principle 'Weak Desire Reflection'. It is this weaker principle that he ends up defending.

special case.[71] Now, similarly to what we acknowledged in the belief case, it could be that you predict an experience by which your awareness grows to be so 'personally transformative' (Paul 2014) that your fundamental values will change in such a way that you will (metaphorically speaking) not be the same person. Experimenting with planetary-scale geoengineering or with hallucinogenic drugs might be transformative in this way, for the global community and for a single person respectively. Again, we do not want to rule out the possibility that, upon awareness growth, your values change in such a way that you will not (metaphorically speaking) be 'the same person', or, to put it less dramatically, that you are not stable throughout the experience of awareness growth. But we nevertheless set such possibilities aside for now. And note that the principle does not apply to such cases, since in them it is not true that the *only* thing that happens is that you gain more awareness.

So, let's return our focus to the presumably more typical, non-transformative cases of anticipated awareness growth. Can we make any positive argument in favour of Preference Awareness Reflection? Now, one might think that we already have given an argument for Preference Awareness Reflection. After all, if one satisfies Awareness Reflection (which, we have argued, one should satisfy) and moreover predicts that one's preferences between sure (i.e., risk-free) outcomes won't be affected by the awareness growth, then one will satisfy Preference Awareness Reflection.[72] But that simply raises the question of why or when one should predict one's preferences between sure outcomes to be unaffected by awareness growth. (In particular, even if one has *not* in fact had a transformative awareness growth, can one's preferences between sure outcomes nonetheless rationally change?) Rather than taking for granted that such preferences should be unaffected by (non-transformative) awareness growth, we will argue for Preference Awareness Reflection directly, which has implications for when one's preferences between sure outcomes should be unaffected by awareness growth.

Unlike the Awareness Reflection principle for belief, a traditional (diachronic) Dutch book argument, like that discussed in Section 7.3.2, cannot be made against you if you violate Preference Awareness Reflection. However,

---

[71] For instance, Hedden (2015) argues that what we called Preference Reflection is undermined by the fact that what you prefer often depends on what you have chosen; hence, Preference Reflection will often imply that what you now should prefer depends on what you believe you will choose. However, if between now and time *t* you have made some choice, then Preference Awareness Reflection will not hold, since in that case it is not true that 'between now and time *t*, the only thing that happens is that you gain more awareness'. Similarly with personal identity: if your identity changes between now and *t*, then it is not true that you have only gained awareness between now and *t*. And so on.

[72] We thank Brian Hedden for pressing us on this point.

you are vulnerable to a more general 'dynamic consistency' argument. In particular, you will be willing to pay a price to limit your future options and to bind yourself if your preferences violate Preference Awareness Reflection; so, having preferences that violate this principle may be costly.

For instance, suppose that while you now strictly prefer going to the beach to staying home, you predict that you will prefer to stay home if you become aware of the content of the catch-all weather contingency. Then there is some price that you should be willing to pay to remove the option of staying home if and when you have become aware of the catch-all. Similarly, you should be willing to accept some cost to make a binding decision now to go to the beach.[73] And one might think that there is something irrational about an attitudinal state that makes one vulnerable to such unfortunate, and seemingly unnecessary, expenses.

However, one might wonder whether the dynamic consistency argument doesn't prove much. One can often predict, at some stage of one's life, that one will have different preferences at later stages in one's lives. For instance, when the sleepless nights that are associated with raising toddlers are fresh in one's memory, one might undergo a vasectomy to prevent oneself from acting on a future temptation to have another child. But it would seem that paying to limit one's options can in that case be perfectly rational. Is the predicted preference change in the beach-or-home example, and the associated willingness to bind oneself in that case, any different?

One potential difference, that might suggest that the dynamic consistency argument does at least indicate some problem in the beach-or-home example, is that in that example one expects to undergo a preference change as a result of *gaining more information or worldly wisdom*. So, the person who pays to bind herself from staying at home if she becomes more aware is binding herself from acting on more information than what she now has. In other words, she is accepting a price for being able to act on less rather than more information. And that does seem irrational, at least if one assumes that the gain in information – that is, the awareness growth – is not associated with some fundamental transformation (which seems plausible to assume in the beach-or-home example).[74] In contrast, in the vasectomy case, binding may seem

---

[73] The same goes, of course, for outcomes: if you expect that your preference between two outcomes will change when your awareness grows, then you might now be willing to pay to bind yourself from swapping outcomes in the event that your awareness grows.

[74] Note that in this respect the dynamic argument for Preference Awareness Reflection is like the much discussed dynamic arguments in favour of the Independence axiom and the Sure Thing principle. For a discussion, see e.g. McClennen (1990). See too Steele (2010, 2018) for general discussions of dynamic arguments for principles of rationality.

rational because we assume that the person has more information, or a more vivid memory of what it is like to have small children, when he makes the decision compared to some later time. Alternatively, he may predict that he will undergo radical transformation as he gets older. In any case, paying a price to prevent oneself from acting on a more aware epistemic state could be irrational in a way that many instances of binding are not.

7.6 Concluding Remarks on Section 7

Let's take stock of the main message of this section. We have considered very strict requirements on agents who anticipate awareness growth; namely, requirements that such agents neither expect that their credences be affected on balance by awareness growth nor predict (in the sense of footnote 68) that awareness growth will reverse any of their preferences.

This finding provides an additional reason for including a 'catch-all' when modelling an agent who anticipates awareness growth. Suppose that an agent is choosing between options f and g, abstractly represented in Table 15, and let's assume that the agent reasons that conditional on all the events that she is aware of – that is, events E_1 to E_n – she would prefer f to g; nevertheless, she unconditionally prefers g to f. Given the above argument, we cannot rationalise this preference pattern by stipulating that the agent's expectation of expected utility for f and/or g upon awareness growth diverges from her current expected utility for f and/or g. Such divergence would be irrational. Instead, we can rationalise it by stipulating that the probability weighted utility of the catch-all outcome for f is sufficiently negative, compared to the catch-all outcome for g, that the overall expected utility of g is greater than that of f. But then we need to include a catch-all when modelling this agent's epistemic state. And this catch-all should of course be what we called 'subjective'; that is, standing in for possibilities that the agent herself thinks she is missing and may later become aware of, rather than, say, standing in for the set of all possibilities that she has actually left out.

8 Conclusion

8.1 'Whereof One Cannot Speak, Thereof One Must Be Silent'

We noted at the outset of this Element that there are surely limits to what can be said about one's own limited awareness at a time. In reasoning, one tries to account for all the possible ways that the world might be that are relevant to one's practical purposes. But this reasoning is limited by one's vantage point. One may not be able to discern all the possible contingencies that an onlooker or even one's later self is able to discern. Those things of which one is unaware are forcibly absent from one's reasoning.[75]

The question is whether limited awareness, duly recognised as an absence in an agent's reasoning, may nonetheless play a role in her reasoning. This Element has built on previous work by others in reckoning with this key question. By way of analogy: even if a driver is not able to see potholes in the road, she may react in better and worse ways when she encounters a pothole, and she may practice better and worse defensive driving to avoid or guard against any fallout from such encounters. Similarly, even if one is not aware of all relevant contingencies, perhaps one may adjust one's reasoning in better and worse ways when one encounters a contingency of which one was previously not aware, and perhaps there are better and worse ways to reason defensively in order to guard against any fallout from such encounters.

Standard decision theory does not deal in such reasoning 'potholes' and the normative issues they raise. It simply ignores them, assuming that the contingencies of which an agent is aware at a time are just those of which she is ever aware, at any time. The problem is that this does not do justice to the experiences of many in reasoning. We do apparently encounter our own limited awareness as we undergo awareness growth. And this leads us to expect and be wary of encountering awareness growth in similar kinds of scenarios. Standard decision theory thus seems to let us down. We would ideally like to use decision theory to help us reason about what to do about the world's most pressing problems, such as, say, climate change and species extinction. However, as the examples throughout this Element suggest, these are the types of decisions where we have particularly strong reasons to think that our choices may result in outcomes, or depend on contingencies, of which we are currently unaware. Similarly, we might, in these important situations, suspect that there are options of which we are currently unaware. Since the standard decision-models, such as those of Savage (1954) and Jeffrey (1965), were developed for decision-making with full awareness, they are apparently not well-suited to help us solve these pressing problems.

Still, wish as we might for a way to deal with reasoning potholes, especially when the stakes are high, they may evade norms of rationality. Trying to reason *beyond uncertainty* may be akin to searching for the holy grail. While this Element joins the search for norms of rationality for responding to as well as anticipating growing awareness, we do not presume that there are any such

---

[75] Hence the use of a well-known quote from Wittgenstein (1922, proposition 7) as the title of this section. While apt (roughly speaking) for our purposes, the quote conveys, in Wittgenstein's discussion, a rather different point.

norms to be found. Indeed, the conclusions that we arrive at over the course of the Element are somewhat ambiguous: we suggest norms for responding to and anticipating growing awareness, but we also cast doubt on whether these norms are truly substantive or distinct. Even when read in the most deflationary way, however, our analysis shows that reckoning with limited awareness is important for finessing one's reasoning roadmap at any given time – for determining what, all things considered, is one's best assessment of the possible contingencies, given the various learning experiences and further choices one may encounter later.

8.2 Norms for Limited Awareness

Let us then recap the normative conclusions of the Element, however substantive or distinct one may regard them.

Note that an initial task, in order to even begin contemplating norms, was to characterise limited awareness and subsequent awareness growth. Hence in Section 2 we described decision models that represent an agent's own fallible perspective. Such models may fail to account for all possible contingencies, and when/if an agent recognises this, she undergoes awareness growth. In Section 3 we considered more carefully how to model a transition from one state of awareness – what we dub an *awareness context* – to another. We suggested that this modelling exercise is in itself enlightening with respect to better understanding an agent's reasoning at a time. Indeed, we hope that our discussion in these early sections not only contributes to the awareness literature but more generally highlights the limits of the standard possible-world models of cognitive states.

With a general model of awareness growth in hand, we were in a position to investigate potential norms of rationality for, respectively, responding to and anticipating awareness growth. We approached these topics in turn, the former in Sections 4 and 5 and the latter in Sections 6 and 7.

One might say that the most striking new norm(s) we identified regarding limited awareness were those canvassed at the end of the Element: awareness version(s) of the so-called *Reflection Principle*. The belief norm, which we dubbed *Awareness Reflection*, applies in cases where an agent anticipates her awareness growth in a rather precise way: she can specify all her possible future credences (over those propositions in her current awareness context) after her awareness has grown, whatever it is that she comes to be aware of. If, in addition, the agent takes herself to be the 'same epistemic agent' after the awareness growth in question, her current credences should be the expectation of her future credences.

We take Awareness Reflection to be a substantive and distinct belief norm. There are reasons why it might be downplayed. In particular, while on first glance the norm appears to strongly constrain how an agent's credences at one time relate to her credences at another time, on closer inspection one sees that it is not actual future credences but rather predicted future credences that play a role. Once this is appreciated, the norm can be understood as simply describing how a rational agent arrives at her current credences: by thinking through what are her possible better-informed credences and taking the expectation of these future credences. This is very much in line with the familiar Reflection Principle, and indeed in Sections 6 and 7 we emphasised the continuity between 'ordinary' reasoning and reasoning that involves anticipated awareness growth. But that continuity is the very reason we regard Awareness Reflection to be a surprising norm. The fact that anticipated awareness growth is akin to anticipated learning of a more ordinary kind is itself an important finding.

Our analysis of anticipated awareness growth brings further perspective to the earlier, more negative findings of the Element. Initially we set out to explore simply whether there are better and worse ways to respond to some particular awareness growth (or alternatively whether there are ways to discern whether credences are 'stable' upon awareness growth). In other words, does rationality (or stability) impose any general constraints on the relationship between one's credences prior to and post some growth in awareness? Against the popular position in the philosophy and economics literature, we argued that there are no such general constraints. This position can be seen to resonate with Awareness Reflection in that this norm does not constrain the predictions that an agent makes about her future credences after awareness growth. An agent may predict any kind of credence change, including rather radical changes, so long as her current credences are the expectation of her predicted future credences post awareness growth.

Nevertheless some of our examples in Sections 4 and 5 suggest there will be many occasions in which an agent's credence change upon awareness growth will be more minimal. These are interesting cases to characterise. We do so in the form of our *Restricted Reverse Bayesianism* rule. We do not take the rule to be a substantive norm; it simply describes cases in which what the agent becomes aware of is *evidentially irrelevant* to the pair of 'old' propositions that is of interest. In such circumstances, it follows that the relevant aspect of the agent's belief change will be conservative in the familiar way: the relative probabilities of the pair of propositions in question will remain constant. We suggest that others who have proposed more general norms for conservative belief change under awareness growth either explicitly or implicitly focus only on cases in which the awareness growth is evidentially irrelevant to the pair of propositions at issue.

8.3 Two Challenges Revisited

All this speaks to at least one of the challenges that we raised in the introductory section with respect to whether studying limited awareness and awareness growth is a worthwhile project. One of the worries was whether there could be any normative upshots from better understanding limited awareness, since, compared to, say, failures of transitivity, a lack of awareness is not something that an agent can change through reasoning alone. We suggested that the proof would be in the pudding, and we hope that our summary of the Element's findings in the last section makes for a convincing case that our project is indeed interesting from a normative perspective. At the very least, the process of modelling limited awareness and awareness growth is important for a more sophisticated view of reasoning at a time.

The other challenge raised in Section 1 concerns whether a study of limited awareness could be scientifically respectable. That is, the worry was that, by introducing lack of awareness into a model of an agent, we would inevitably have to start making assumptions about agent that cannot even in principle be empirically verified.

One partial response to this challenge, which we briefly discussed in Sections 6 and 7, was that as long as an agent has a consistent *conditional* preference relation, given all the events of which she is aware, we can infer a great deal about her attitudes to that which she takes herself to be unaware of from her all-things-considered preference (and, ideally, choice) between options for which she is open to, and may even anticipate, awareness growth. Moreover, as we mentioned in the introduction, decision theories have already been developed that allow for a representation of agents' attitudes to that of which they are unaware.

However, for the readers not convinced by these responses, we can offer a partner-in-crime response. As we discussed in the introduction, we see the extension to limited awareness as a natural next step in decision theory's historical trajectory, from the *objective* expected utility theory of von Neumann and Morgenstern (1947), where the only subjective element is the extent to which agents desire outcomes, to the *subjective* expected utility theories of Savage (1954) and Jeffrey (1965), where the extent to which agents believe that various events will occur is also subjective. As we pointed out, the latter two theories however do not allow for any subjectivity when it comes to what is possible or available – that is, they do not allow for limited awareness.

Now, a well-known problem – or feature, depending on one's philosophical views – of introducing the additional subjective variable to represent agents' beliefs is that it invariably introduces some (additional) assumptions about agents that cannot even in principle be empirically verified. For instance, Savage assumes that agents have preferences between any functions from his set of states of the world to his set of consequences. But, as has been much discussed, some of these functions will correspond to options that are not only physically (perhaps even *meta*physically) impossible but also impossible according to the agent whose attitudes are being represented (see e.g. Joyce 1999: chapter 3). And evidently we cannot devise a choice scenario that reveals an agent's preference between options that she thinks are impossible.

Jeffrey on the other hand does not make as strong non-empirical assumptions as Savage. But an implication of this is that Jeffrey's framework results in inconsistent representations of agents' attitudes. In particular, for any agent who satisfies all assumptions of Jeffrey's framework, and for most[76] contingent but logically independent propositions A and B, the agent will both be representable as believing A more strongly than B and B more strongly than A.[77] However, since people who satisfy all of Jeffrey's assumptions have to be very rational indeed, it is assumed that this seeming inconsistency is a problem with the representation, not with the agents themselves. But this assumption cannot, within Jeffrey's system, be empirically justified – what can at best be observed, namely, an agent's preferences as revealed by her choices, is consistent with her both believing A more strongly than B and B more strongly than A.

In sum, the move from objective to subjective expected utility theory, which was celebrated as a great achievement, brought with it assumptions about agents that are even in principle empirically unverifiable. So, while the hard-lined empiricist may still not be comforted by this, at least we take comfort in the fact that we are in good company in accepting the introduction of empirically unverifiable assumptions about agents as a cost of making decision theory more subjective.

8.4 Connection to Applied Work

Alongside theoretical developments in modelling and understanding limited awareness and growing awareness, there has and continues to be progress of a more practical kind. This is in the form of decision support tools for relatively

---

[76] In particular, for all propositions that are not of 'neutral' desirability, or very close to neutral desirability; that is, for all propositions that are not close to the desirability of the tautology.

[77] This assumes that the agent's preferences are not unbounded. For a discussion of this assumption, see Joyce (1999: chapter 4).

novel and/or complex decisions that assist in identifying what are the relevant possible contingencies and one's attitudes towards them. We regard this work as complementary to the more general and abstract treatment of growing awareness that is the focus of this Element. Ideally the two would inform each other.

For instance, one popular family of decision-support approaches is known as *scenario-based planning* (Schwartz 1996). The aim is to assist in mapping out the range and boundaries of the possibility space for a given decision – to determine just how disparate are the scenarios or fully detailed ways that the world might be that are pertinent to the choice at hand. A simple qualitative approach to this effect involves identifying the key factors or 'axes' that discriminate the outcomes of different options. For instance, two such key factors with respect to future global emissions scenarios (that are pertinent to mitigation decisions) are purportedly demographic (population) change and the rate and direction of technological change (Nakicenovic et al. 2000). A telling range of emissions scenarios can thus be constructed from combinations of extreme values on these and the other key axes. In other cases, the key axes and associated spread of scenarios may be less easy to discern with the naked eye, so to speak. If it so happens that there is nonetheless a rich predictive model available, computer-assisted scenario discovery approaches (e.g., Groves and Lempert 2007, Bryant and Lempert 2010) may be useful. One such approach involves automated identification and clustering of what may be hundreds to millions of potentially important scenarios generated by a complex predictive model with large ranges for the parameter values. The idea is that once scenarios are clustered they are cognitively accessible and thus more meaningfully evaluated.

We suggest that scenario-based planning and other decision-support approaches associated with 'horizon scanning' are, among other things, implicitly techniques for regimenting limited awareness and anticipated awareness growth, in that they aim for coverage of the possibility space even if all the 'interior' possibilities cannot be specified in full detail. That is, these decision-support approaches may be usefully conceived as techniques for overcoming limited awareness, such that any anticipated future awareness growth is by refinement – and moreover refinement that is evidentially irrelevant to the comparison of propositions already known, such that Reverse Bayesianism holds and credences remain unchanged. Better still if the value of the coarse outcomes will also be unaffected by any future awareness growth, such that the evaluation of the options is not distorted by limited awareness.

As such, these methods may effectively flesh out what reckoning with limited and growing awareness looks like in the context of real and significant decision problems. This more practical task has not been our preoccupation in this Element. But we hope that our work here may provide food for thought and helpful markers in the sand, so to speak, for those working at this important practical end of the spectrum. Reverse Bayesianism is not a requirement of rationality, as we have argued. But a decision-maker may nonetheless aspire to overcome their limited awareness to the point where any potential awareness growth is predicted to be a refinement that more or less satisfies Reverse Bayesianism, leaving all or most relative credences unchanged. Indeed, a decision-maker would presumably further hope that the values of outcomes and ultimately which of the available options is deemed optimal is more or less unaffected by their limited awareness.

Even if that is not the *explicit* aim of the scenario-planning and related decision support methods – presumably the stated aim is simply to recognise all available options and evaluate them as well as possible[78] – we suggest that it is a useful interpretation of the methods. It makes clear the ways in which the methods may reasonably *fail* in any given application. Namely, the methods may fail to overcome limited awareness such that any further anticipated awareness growth has only minor impact on the evaluation of the options. It may turn out, for instance, that the anticipated awareness growth has more dramatic effects on the relative expected utilities of options than predicted or else that awareness turns out to grow in altogether unanticipated ways.

8.5 Further Research

When we introduced the topic of our inquiry in this Element – limited awareness and how it changes over time – we appealed to an agent's plight in making decisions about what to do in the world. We made clear that our very understanding of an agent's epistemic state, in addition to her preferences, relates to the decision problem she faces at a given time. Hence we position our project within a larger decision-theory narrative. And yet, throughout the Element, many of our examples (particularly those in Sections 4 and 5) focus just on one aspect of an agent's decision problem – the ways the world might be that affect the outcomes of her options, or the *state space*. Our paradigm for awareness growth is thus expansion or refinement of the state space. We acknowledged in various places that limited awareness and subsequent awareness growth may also affect other aspects of an agent's decision problem. In particular, the agent may have limited awareness with respect to the set of options available to her,

---

[78] In fact, some of these methods eschew probabilistic and expected utility reasoning altogether, opting for less informationally demanding ways of evaluating options in light of their possible outcomes.

and/or she might have limited awareness with respect to the outcomes that will result from particular options in particular states of the world.

Our presumption has been that that limited awareness and awareness growth with respect to the option and outcome aspects of an agent's decision problem may be treated just as per the state space. After all, options and outcomes are propositions in our (Jeffrey-inspired) model of the agent's reasoning, just like states are. So in all cases it is simply a matter of the agent's underlying proposition space changing, whether by expansion or refinement. Moreover, it is important to note that by studying (un)awareness of states we have at least implicitly studied (un)awareness of consequences too. A discovery of a new state often introduces a new consequence, such as in the rent-or-buy example, from Section 3, where the discovery of the possibility that the landlord sells the apartment (i.e., the discovery of a state) introduces the possibility of becoming homeless (i.e., an outcome) which was not part of the original decision problem.

The idea that all propositions in a decision model (states, outcomes and options) can be given the same treatment, as far as (un)awareness is concerned, is, we suggest, a suitable starting position. Nevertheless, it deserves further examination. In particular, consider limited awareness about the option space. On the one hand, options are simply propositions, and, just as an agent may revise her state space, she may later become aware that she has (or had) more options at a given time than she had realised, or that the options at hand are (or were) more refined than she had realised. And yet, on the other hand, perhaps there are further interesting questions in connection with the option space. These are not just any old propositions; they are propositions that describe, by the agent's own lights, things she can will to happen in the world. Should we expect that special puzzles arise in connection with awareness growth for propositions of this sort? That is an important question for future work.

Another avenue for further research concerns the potentially idealised nature of our basic model for studying limited awareness and awareness growth. As noted at the outset, our methodology in studying limited awareness was to consider this phenomenon in isolation. We thus set aside various other challenges to orthodox expected utility theory. For instance, we did not entertain generalisations of expected utility theory that accommodate alternative representations of an agent's uncertainty such as imprecise probabilities (or sets of probability functions). Nor did we entertain generalisations of expected utility theory that accommodate, on the preference side, more complex kinds of risk aversion. For those sympathetic to theories of either kind, our analysis can be understood as making certain idealising assumptions, for instance, that in the cases we examine the agent's credences are precise probabilities and she is moreover risk-neutral in whatever is the relevant sense of that term.

But idealising in this way may not be apt if in fact the phenomenon of awareness growth is not independent (in a normative sense) of the representation of uncertainty and/or risk aversion. Let us focus just on the former issue. One might think that an appropriate response to awareness growth requires 'going imprecise': roughly, the 'new' propositions of which one has become aware are assigned maximally imprecise credence (perhaps conditional on some relevant 'old' proposition). Moreover, one might think that when an agent anticipates awareness growth, her uncertainty is often so severe that it is best represented by imprecise probabilities. As such, one might argue that allowing credences to be imprecise is crucial for modelling awareness growth and the anticipation of it in a compelling way, and for expressing the relevant norms.

While we have some sympathy for this line, we think it would be very surprising indeed if imprecise credence were rationally mandated for growing awareness and/or its anticipation. And, as noted, we do not think the mere permissibility of imprecise credence in contexts of growing awareness undercuts our analysis. Those who subscribe to imprecise credence being rationally permissible can view our method as one of idealisation: we assume that in all cases under investigation the agent happens to have precise credences. Ditto for non-standard kinds of risk aversion. It would be extremely surprising if growing awareness mandated some particular risk-averse attitude. And if non-standard kinds of risk aversion are merely permissible, then this does not compromise our approach to studying growing awareness.

All that said, it remains an interesting project to relax any idealisations or scope restrictions that are inherent in our analysis. We welcome further investigation of how growing awareness interacts with non-standard representations of uncertainty and risk aversion. Note that some inroads have already been made on this project, at least with respect to imprecise attitudes. For instance, Bradley (2017: chapter 12) permits imprecise attitudes in his exposition of growing awareness. Economists Dominiak and Tserenjigmid (2018) explore in a working paper the transition from precise to imprecise probabilities upon awareness growth and suggest constraints on such a transition. Our findings in this Element provide a base from which to examine these models/proposals that relate growing awareness to imprecise attitudes.

There is a further idealisation in our study of growing awareness that is tangential to the treatment of uncertainty and risk aversion. Our account makes the standard assumption that agents *progress* in their reasoning and view on the world. In particular, no forgetting or becoming less aware is allowed. The assumption of exclusive progress may be relatively harmless for most normative enquiries. However, when it comes to awareness growth, there may be progressive reasons to eradicate old concepts in one's proposition space, since it may not always be a matter of forgetting. Rather, it may be a matter of conceptual learning, where old and outdated concepts are replaced by new ones.

We make this suggestion with caution, because we suspect that this way of thinking of an agent's concepts – they they may be replaced by more explanatory or in some other sense more apt concepts – would represent a radical departure from the notion of reasoning and learning that decision theory has been designed to capture. It would involve seeing learning not as a linear accumulation of knowledge but rather a jerky process that admits of conceptual revolutions à la Kuhn (1962). So this may be very far from incremental future research. And it may ultimately be regarded beyond the scope of decision theory. But all the more reason, we say, to make some preliminary investigations to see how different reasoning would look were we to allow for progressive 'loss of awareness'.

8.6 Closing Remarks

As the (incomplete) list of issues not addressed in this Element illustrate, there is still a lot of work to be done on the connection between rationality and limited awareness. That is, there remain important theoretical questions quite apart from the many further practical ones associated with making good decisions under limited awareness. But we believe that we have nevertheless clarified and moved the frontier of research on limited awareness. Our main hope for this Element, however, is simply that it will bring further attention to the importance of studying limited and growing awareness, this being a fundamental aspect of our predicament in the world as reasoning agents.

Bibliography

Anscombe, F. J. and R. J. Aumann (1963). A definition of subjective probability. *Annals of Mathematical Statistics 34*(1), 199–205.

Arntzenius, F. (2008). No regrets, or: Edith Piaf revamps decision theory. *Erkenntnis 68*(2), 277–97.

Bolker, E. D. (1967). A simultaneous axiomatisation of utility and subjective probability. *Philosophy of Science 34*(4), 333–40.

Bradley, R. (2005). Radical probabilism and Bayesian conditioning. *Philosophy of Science 72*(2), 342–64.

Bradley, R. (2017). *Decision Theory with a Human Face*. Cambridge University Press.

Bradley, S. (2019). Imprecise probabilities. In E. Zalta (ed.), *Stanford Encyclopedia of Philosophy*. https://plato.stanford.edu/entries/imprecise-probabilities/.

Briggs, R. A. (2009). Distorted reflection. *Philosophical Review 118*(1), 59–85.

Briggs, R. A. (2017). Normative theories of rational choice: Expected utility. In E. Zalta (ed.), *Stanford Encyclopedia of Philosophy*. https://plato.stanford.edu/entries/rationality-normative-utility/.

Bryant, B. P. and R. J. Lempert (2010). Thinking inside the box: A participatory, computer-assisted approach to scenario discovery. *Technological Forecasting and Social Change 77*(1), 34–49.

Buchak, L. (2013). *Risk and Rationality*. Oxford University Press.

Bykvist, K. and H. O. Stefánsson (2017). Epistemic transformation and rational choice. *Economics and Philosophy 33*(1), 125–38.

de Canson, C. (ms.). The nature of awareness growth.

Diaconis, P. and S. L. Zabell (1982). Updating subjective probability. *Journal of the American Statistical Association 77*(380), 822–30.

Dominiak, A. and G. Tserenjigmid (2018). Ambiguity under growing awareness. Available at SSRN: https://ssrn.com/abstract=3247761 or http://dx.doi.org/10.2139/ssrn.3247761.

Earman, J. (1992). *Bayes or Bust? A Critical Examination of Bayesian Confirmation Theory*. MIT Press.

Fagin, R. and J. Y. Halpern (1987). Belief, awareness, and limited reasoning. *Artificial Intelligence 34*(1), 39–76.

Gilboa, I. and D. Schmeidler (1995). Case-based decision theory. *Quarterly Journal of Economics 110*(3), 605–39.

Gilboa, I. and D. Schmeidler (2001). *A Theory of Case-Based Decisions.* Cambridge University Press.

Glymour, C. (1980). Why I am not a Bayesian. In C. Glymour (ed.), *Theory and Evidence*, pp. 63–93. Princeton University Press.

Grant, S. and J. Quiggin (2013a). Bounded awareness, heuristics and the precautionary principle. *Journal of Economic Behavior & Organization 93*(C), 17–31.

Grant, S. and J. Quiggin (2013b). Inductive reasoning about unawareness. *Economic Theory 54*(3), 717–55.

Groves, D. G. and R. J. Lempert (2007). A new analytic method for finding policy-relevant scenarios. *Global Environmental Change 1 7*(1), 73–85.

Hájek, A. (2003). What conditional probability could not be. *Synthese 137*(3), 273–323.

Hájek, A. Omega. Unpublished manuscript.

Hansson, S. O. (2009). From the casino to the jungle. *Synthese 168*(3), 423–32.

Hedden, B. (2015). *Reasons Without Persons: Rationality, Identity, and Time.* Oxford University Press.

Heifetz, A., M. Meier, and B. Schipper (2006). Interactive unawareness. *Journal of Economic Theory 130*(1), 78–94.

Heifetz, A., M. Meier, and B. C. Schipper (2008). A canonical model for interactive unawareness. *Games and Economic Behavior 62*(1), 304–24.

Henderson, L., N. D. Goodman, J. B. Tenenbaum, and J. F. Woodward (2010). The structure and dynamics of scientific theories: A hierarchical Bayesian perspective. *Philosophy of Science 77*(2), 172–200.

Hill, B. (2010). Awareness dynamics. *Journal of Philosophical Logic 39*(2), 113–37.

Jeffrey, R. (1965). *The Logic of Decision.* University of Chicago Press.

Joyce, J. M. (1998). A nonpragmatic vindication of progabilism. *Philosophy of Science 65*(4), 575–603.

Joyce, J. M. (1999). *The Foundations of Causal Decision Theory.* Cambridge University Press.

Karni, E. and M.-L. Vierø (2013). 'Reverse Bayesianism': A choice-based theory of growing awareness. *American Economic Review 103*(7), 2790–810.

Karni, E. and M.-L. Vierø (2015). Probabilistic sophistication and reverse Bayesianism. *Journal of Risk and Uncertainty 50*(3), 189–208.

Karni, E. and M.-L. Vierø (2017). Awareness of unawareness: A theory of decision making in the face of ignorance. *Journal of Economic Theory 168*, 301–25.

Kuhn, T. S. (1962). *The Structure of Scientific Revolutions.* University of Chicago Press.

Maher, P. (1995). Probabilities for new theories. *Philosophical Studies 77*(1), 103–15.

Mahtani, A. (2020). Awareness growth and dispositional attitudes. *Synthese*, 1–17.

McClennen, E. F. (1990). *Rationality and Dynamic Choice: Foundational Explorations*. Cambridge University Press.

Nakicenovic, N., et al. (2000). *Special report on emissions scenarios: A special report of Working Group III of the Intergovernmental Panel on Climate Change*. Cambridge University Press.

Paul, L. A. (2014). *Transformative Experience*. Oxford University Press.

Pettigrew, R. (2020). *Dutch Book Arguments*. Cambridge University Press.

Piermont, E. (2017). Introspective unawareness and observable choice. *Games and Economic Behavior 106*(C), 134–52.

Quiggin, J. (2016). The value of information and the value of awareness. *Theory and Decision 80*(2), 167–85.

Ramsey, F. P. (1990/1926). Truth and probability. In D. H. Mellor (ed.), *Philosophical Papers*, pp. 52–94. Cambridge University Press.

Roussos, J. (2020). *Policymaking under scientific uncertainty*. PhD thesis, London School of Economics and Political Science.

Savage, L. (1954). *The Foundations of Statistics*. John Wiley & Sons.

Schipper, B. C. (2014). Awareness. http://faculty.econ.ucdavis.edu/faculty/schipper/unawhb.pdf.

Schipper, B. C. (2015). Awareness. In H. van Ditmarsch, J. Y. Halpern, W. van der Hoek, and B. Kooi (eds.), *Handbook of Epistemic Logic*, pp. 77–146. College Publications.

Schwartz, P. (1996). *The Art of the Long View: Planning in an Uncertain World*. Currency-Doubleday.

Shimony, A. (1970). Scientific inference. In R. Colodny (ed.), *The Nature and Function of Scientific Theories*, pp. 79–172. University of Pittsburgh Press.

Stalnaker, R. (1984). *Inquiry*. MIT Press.

Steele, K. (2010). What are the minimal requirements of rational choice? Arguments from the sequential-decision setting. *Theory and Decision 68*(4), 463–87.

Steele, K. (2018). Dynamic decision theory. In S. O. Hansson and V. F. Hendricks (eds.), *Introduction to Formal Philosophy*, pp. 657–67. Springer.

Steele, K. and H. O. Stefánsson (2015). Decision theory. In E. Zalta (ed.), *Stanford Encyclopedia of Philosophy*. https://plato.stanford.edu/entries/decision-theory/.

Stefánsson, H. O. and R. Bradley (2019). What is risk aversion? *British Journal for the Philosophy of Science 70*(1), 77–102.

Titelbaum, M. G. (2012). *Quitting Certainties: A Bayesian Framework Modeling Degrees of Belief.* Oxford University Press.

Vallinder, A. (2018). *Bayesian variations: essays on the structure, object, and dynamics of Credence.* PhD thesis, London School of Economics and Political Science.

van Fraassen, B. C. (1984). Belief and the will. *Journal of Philosophy 81*(5), 235–56.

Vineberg, S. (2011). Dutch book arguments. In E. Zalta (ed.), *Stanford Encyclopedia of Philosophy.*

von Neumann, J. and O. Morgenstern (1947). *Games and Economic Behavior* (2nd edn). Princeton University Press.

Walker, O. and S. Dietz (2011). A representation result for choice under conscious unawareness. Grantham Research Institute on Climate Change and the Environment Working Paper No. 59.

Wenmackers, S. and J. Romeijn (2016). New theory about old evidence. *Synthese 193*(4), 1225–50.

Wittgenstein, L. (1922). *Tractatus Logico-Philosophicus.* Kegan Paul.

Zabell, S. L. (1992). Predicting the unpredictable. *Synthese 90*(2), 205–32.

Acknowledgements

We have benefited greatly from discussing the topic of this Element – and in many cases, draft sections – with a number of people, including Christian Barry, Richard Bradley, Kamilla Haworth Buchter, Krister Bykvist, Catrin Campbell-Moore, Chloé de Canson, Edward Elliott, Simon Grant, Hilary Greaves, Thor Grünbaum, Alan Hájek, Sven Ove Hansson, Brian Hedden, Casey Helgeson, Anna Mahtani, Andreas Mogensen, Michael Nielsen, Ignacio Ojea Quintana, Wlodek Rabinowicz, Joe Roussos, Julia Staffel, Daniel Stoljar, Christian Tarsney, Benjamin Tereick, Teruji Thomas, Aron Vallinder, Marie-Louise Vierø, and two referees for Cambridge University Press. We acknowledge too the significant role played by Martin Peterson, the editor for this Cambridge series, in making this Element happen. Note that parts of Sections 3, 4 and 5 are based on our forthcoming article 'Belief Revision for Growing Awareness', on which we received very useful comments from the editors and anonymous referees of *Mind*.

We also received helpful comment when presenting work related to this Element at the 'Workshop for the Beyond Rational Choice Project', Australian National University, 2016; the Higher Seminar in Philosophy, Umeå University, 2017; the 'Stockholm Workshop on Philosophy & Economics'; the KTH Royal Institute of Technology, 2017; 'What are Degrees of Belief?', University of Leeds, 2018; the 'Foundations of Normative Decision Theory' workshop, University of Oxford, 2018; the 'Ethics and Risk' workshop, Australian National University, 2018; the 'Epistemic and Personal Transformation' workshop, University of Queensland, 2019; a reading group at the Global Priorities Institute, University of Oxford, 2020; and the Joint Stockholm/Uppsala Seminar in the Philosophy of Science, Stockholm University, 2020.

Orri gratefully acknowledges financial support from Riksbankens Jubileumsfond through a Pro Futura Scientia XIII fellowship. Katie's research was supported by an Australian National University (ANU) Futures grant, an Australian Research Council Discovery grant (grant number 170101394) and the Humanising Machine Intelligence 'Grand Challenge' project at ANU. Finally, both Katie and Orri have received financial support from the project 'Climate Ethics and Future Generations' at the Institute for Futures Studies in Stockholm, which is funded by Riksbankens Jubileumsfond (grant number M17-0372:1).

Cambridge Elements ≡

Decision Theory and Philosophy

Martin Peterson
Texas A&M University

Martin Peterson is Professor of Philosophy and Sue and Harry E. Bovay Professor of the History and Ethics of Professional Engineering at Texas A&M University. He is the author of four books and one edited collection, as well as many articles on decision theory, ethics and philosophy of science.

About the Series

This Cambridge Elements series offers an extensive overview of decision theory in its many and varied forms. Distinguished authors provide an up-to-date summary of the results of current research in their fields and give their own take on what they believe are the most significant debates influencing research, drawing original conclusions.

Cambridge Elements ≡

Decision Theory and Philosophy

Elements in the Series

Dutch Book Arguments
Richard Pettigrew

Rational Choice Using Imprecise Probabilities and Utilities
Paul Weirich

Beyond Uncertainty: Reasoning with Unknown Possibilities
Katie Steele and H. Orri Stefánsson

A full series listing is available at: www.cambridge.org/EDTP

Printed in the United States
by Baker & Taylor Publisher Services